THE MAELSTROM WHIRLPOOL

Written by

MITCH RUBMAN

I0406025

mwhirlpool@gmail.com
mrubman@hotmail.com
323 377 8298
(C)Mitch Rubman
Reading

FADE IN:

EXT. PACIFIC OCEAN - FREIGHTER - NIGHT - LONG SHOT

An enormous Chinese freighter filled with cargo cuts through the Pacific Ocean unbeknownst of its' fate. A Fog bank rapidly approaches. They enter the Fog bank. The air is thick and the freighters windows fill with moisture in the fog. A loud fog HORN sounds.

INT. PACIFIC OCEAN - FREIGHTER - NIGHT

Inside the bridge, the Captain, crew, and first mate, are celebrating New Year's Eve 2010, laughing and toasting. We start to hear a roar from outside. The roar gets louder. The compass spins, gauges go on and off. Lights go and off.

The freighter tilts forty five degrees to one side, throwing everything that isn't tied down all about the cabin. Loud emergency warning signals activate. The sounds of the hull bending are deafening. The freighter moans like a wounded animal. The boat dips forward and then backward. The roar of the water gets louder.

Through the window we see a tall wall of black ocean water spinning by in front of us.

The crew looks at each other helpless as the freighter gets thrown about like a small toy in the massive Whirlpool.

As the sounds of the Whirl getting louder and louder we follow the boat and crew downward into the dark abyss, Spinning downward.

Then the Maelstrom Whirlpool stops. The ocean quickly closes up on the freighter. The water becomes still. The cargo floats upward. The fog disappears. All is quiet. Back to a silent dark night on the Pacific.

EXT. HIDDEN POINT ROAD - ANNAPOLIS, MD. - NIGHT

Black asphalt road surrounded by water on both sides. Sign posts indicate four miles from the Naval Academy.

EXT. ATLANTIC OCEAN - UNDERWATER - RESCUE SUBMARINE PROTOTYPE - CONTINUOUS

We see a sleek black sea turtle like submarine speed through the channel in the inlet. The vessel is towing a torpedo shaped object.

INT. RESCUE SUBMARINE PROTOTYPE - NIGHT

We dissolve into the dark interior of the sub and see the outline of CAPTAIN JOHN CUTTER, 30s, trying to steer the small submarine. A flash of indicator lights reading "Danger, electrical failure." Cutter slams the control board and all the lights go out.

                    CUTTER
          Huh.

A light goes on, it spells, "EMERGENCY WARNING." Cutter is illuminated by the red blinking light.

Cutter sees depth gauge read 40 feet. Oxygen remaining, one minute.

Puts on a wet suit very quickly. Has three seconds left on the gauge.

Hits switch, "Emergency Air lock release"

EXT. RESCUE SUBMARINE PROTOTYPE

We see Cutter blow the hatch and exit the submarine and swim over to the vessel in tow.

He opens the hatch and enters the rescue vehicle. A burst of air exits.

It is a tighter fit than the other submarine. The power switches on and he's off.

EXT. HIDDEN POINT ROAD - ANNAPOLIS, MD. - CONTINUOUS

Two armed military policeman stand in the rain on the road with binoculars looking at the ocean inlet. We see a ripple coming towards the shore.

The ripple turns to white water, larger and larger pools of foam form. Then a small explosion. We see a dark silhouette of Captain Cutter rise from the water. He walks towards the military police vehicle slowly taking off equipment.

INT. MILITARY POLICE VEHICLE - NIGHT

Cutter gets inside the dark black sedan with two MILITARY POLICE, 20s. We now get to see his face. He is unshaven with a scar running on his right side--slightly. He slips off the wet suit and hands two black envelopes to the MP. He zips up a pair of leather boots.

                    MP
          Glad we could help, Sir.

                    CUTTER
          Prototypes. Still raining.

Cutter changes his shirt. We see scars and bullet wounds on his skin. A holster.

                    MP
          Just last three weeks. Maybe
          you'll bring the sun, Sir.

                    CUTTER
          No question.

The vehicle stops. The MP exits quickly and opens the door with an umbrella in hand.

Cutter steps out of the black sedan.

                    CUTTER (cont'd)
          Don't worry about the umbrella.

Cutter steps out of the vehicle into the rain. Looks up at
the sky, the rain splashing on his face.

EXT. CUTTER'S APT. BUILDING - WASHINGTON D.C. - NIGHT

Outside of his four floor brownstone. His doorway is blocked
by rolled up newspapers. Cutter walks over to his mailbox
which is crammed with mail. There is a bill taped to the door
from the Newspaper Delivery Company. We can see the word
CANCELLED bleeding though the envelope. He opens the door and
walks inside.

INT. CUTTER'S APT. BUILDING - WASHINGTON D.C. - NIGHT

He continues up the stairs to the top floor.

Cutter's brownstone is filled with Navy commendations and
proclamations. There are many pieces of submarines and boats
all over the apartment. In the center is a large submarine
model.

There is a new BLACK SUIT hanging from the top of the
periscope. Price tags still attached.

Cutter goes over to an answering machine. Hits play.

The FIRST message. Machine announces, "nine thirty-one."

                    KITTY (ON MACHINE)
          Just wondering where you
          are...it's already nine thirty.
          Kitty--in case you forgot. Your
          fiancée. Hello! Call me.

The SECOND message. Machine announces, "nine thirty-four."

                    CUTTER
          Hmm.

                    KITTY (ON MACHINE)
          CUTTER-late! Where are you? Your
          cell is off.

The THIRD message. Machine announces, "nine thirty-nine."

                    KITTY (cont'd)(ON MACHINE)
          Darling...If you don't show up at
          our engagement party you had
          better stay under water, cause...
          I'm coming after you.

Cutter stops the machine.

                    CUTTER
               (out loud)
          Kitty.

He walks over to a small dying rubber tree in the corner of the apartment. Sprays water on it. Takes deep breath, lays down on couch and falls asleep.

EXT. CUTTER'S SUV - MORNING

Cutter drives an enormous SUV. There is a bumper sticker which says, "Submarine Captains do it under water and at attention." The sun is shining.

INT. CUTTER'S SUV - MORNING

Cutter is clean shaven, wearing pressed dress blues.

                    RADIO ANNOUNCER
          ...And with heavy flooding
          expected to continue, the Red
          Cross has been called out to help
          those less fortunate. This
          extreme weather is expected for
          today, tonight and through
          tomorrow. This is the worst
          flooding in over a century. There
          are even dust devils being
          reported as faraway as Brooklyn.

Cutter turns off the radio.

                    CUTTER
          Lovely.

EXT. NAVAL ACADEMY - ANNAPOLIS - MARYLAND - DAY

The Naval Academy, built in the 1930's is a complex of grey cement buildings. The MPs pass Cutter through with a nod of recognition.

He pulls up to the Main Campus Building. Stops the car, grabs a dozen roses from the backseat, adjusts his REGALIA pins and metals in the mirror and steps outside.

With a loud crackle it immediately starts to rain. Cutter grabs an old newspaper from the back seat to use as a cover.

Cutter walks by a group of FEMALE RECRUITS smoking cigarettes under an awning. They recognize him and tease him.

                    RECRUIT
          Marry me...please...no one brings
          me roses anymore...See that's
          romantic.

EXT. NAVAL ACADEMY

They hoot. He waves.

                    CUTTER (V.O.)
          I'm never going to hear the end
          of this. It's red roses she
          likes. (cuts himself) Damn
          thorns.

Walking quickly with the flowers Cutter stops in front of a classroom.

TAKES A DEEP BREATH...then knocks on the door and slowly enters the classroom, dripping wet.

INT. NAVAL ACADEMY - CLASSROOM - DAY

He stands in front of a classroom full of STUDENTS. The room is covered with maps and charts of the oceans, currents and sea life. We first meet...

KITTY HONEYWELL, Ph.D., 30s, Oceanographer, pretty, dressed in a khaki dress, Rolex, Ray-bans, is surprised to see Cutter--there is a moment of disbelief and she gives him a double take. Even squinting a little.

She then continues writing on the board. Cutter moves forward.

                    CUTTER
          Hello, Kitty.

Cutter tries to hand her the roses...she just turns around and continues writing.

                    KITTY
          Well, hello Captain, I don't know
          who. So very nice of you to stop
          by. Been a while. Boat in town?

KITTY writes-

SO445 GLOBAL CLIMATE CHANGE

                    CUTTER
          I know. I'm bad.

Cutter hands her a single rose, she puts it down.

                    KITTY
          Sorry class. We're having a
          lesson on oceanography...Would
          you care to join us?

Kitty pulls down a chart of the oceans and grabs a pointer. Looking at Cutter.

                    CUTTER
          Listen...they called me to duty.

Cutter hands Kitty another rose. She puts the rose down next to the other rose. Kitty then picks up a piece of chalk and writes on the whiteboard, "Engagement Party."

                    KITTY
          What? You didn't volunteer. The
          moon causes the pull. Or maybe
          you were busy.

Cutter hands Kitty another rose. She also places it down.

                    CUTTER
          I'm in the Navy, they don't
          always give you your first
          choice. Let me draw it out for
          you.

He then draws an earth-moon system. One big, one small
circle. Very proud of himself.

                    KITTY
          You missed our engagement
          party...The polar ice caps are
          melting. Haven't you seen the
          numbers?

                    CUTTER
          I'm sorry. But, it's a cycle.

Cutter hands her another rose. She also places it down on the
table.

                    KITTY
          Not these. Captain John Cutter.
          My former fiancée. What causes
          the pull?

                    CUTTER
          Gravity...You look very
          beautiful. Class? Here, come on.
          You're giving me global warming.

Cutter hands her another rose.

The students look at each other confused as what to do next.

                    BOY STUDENT
          Maybe we should leave.

                    KITTY
          No. Here.

She hands him her engagement ring. He's a little surprised
and puts it in his breast pocket.

                    CUTTER
          What? The submarine USS
          Sacramento can go to depths of
          over two thousand feet and attain
          speeds of thirty knots.
          Submarines run day and night Can
          you forgive me?

                    BOY STUDENT
          I forgive you.

                    GIRL STUDENT
          He wants to kiss her.

Everyone is still.

                    CUTTER
          Us submarine Captains...we live
          with danger.

                    BOY STUDENT
          Aren't you scared?

                    CUTTER
          I've kissed her before.

                    GIRL STUDENT
          Do you go on his submarine?

                    KITTY
          Only, when we're not GOING out to
          sea.

                    CUTTER
          Kitty. I'm...

Cutter's cell phone rings - A NAVAL JINGLE.

                    CUTTER (cont'd)
          Excuse me.

Cutter hands the rest of the roses to Kitty.

                    KITTY
          Cutter!

                    CUTTER
               (on phone)
          Yes. Sir. (Shakes his head) What?

Cutter gestures for Kitty to go. She looks at him.

                    KITTY
          What did you say? No. Alright.
          I'm going to dismiss the class a
          few minutes early today.

                    CUTTER
          I've got to go.(He interrupts
          her)

Cutter is about to exit. But he stops.

Kitty writes on the board.

                    KITTY
          As I was saying...next week our
          assignment will be to identify...

                    CUTTER
          Let's. (Starts to walk out)

                    KITTY
          ...northern hemispheres and work
          on their mathematical models.

Cutter holds open the door.

INT. NAVAL ACADEMY - HALLWAY - DAY

                    KITTY
          Just who in the world...

The students carefully watch Kitty.

                    CUTTER
          We are to report to NOAA
          headquarters.

                    KITTY
          I am not going with you on any
          sub.

Kitty takes out her cell phone and makes a call. She gives
him a long stare.

                    KITTY (cont'd)
               (on phone)
          Hello? This is Kitty. No. What?
          Well yes. Right. We have to make
          one stop.

They start to exit. Kitty sees the flowers and turns.
Pointing to the girl in the front.

                    KITTY (cont'd)
          Like them?

                    GIRL STUDENT
          Sure. Thanks.

                    KITTY
          Here. Since when do *you* give me
          Roses? Guilty of something, else?

INT. NAVAL ACADEMY - HALLWAY - DAY

They walk through the hallway to the front door. She goes
first.

                    KITTY
          Look, I need to get some of my
          things.

                    CUTTER
          Maybe you shouldn't go.

                    KITTY
          No, I'm going.

Kitty opens the front door with her umbrella. Cutter still
doesn't. The RAIN stops suddenly.

                    KITTY (cont'd)
          Look at that. The rain stopped.
          Glad I don't have abandonment
          issues. I can't tell you what my
          mother said about you or Mitsy.

Cutter reacts to each name mentioned.

EXT. NAVAL ACADEMY - DAY

They walk through the parking area.

                    KITTY
          We're stopping at my house.
          Follow me.

                    CUTTER
          Don't drive fast.

Kitty's Chevy, an environmentally friendly vehicle is parked in front. A big sign on the side of the vehicle states - "Bio-mass and used oil burner." Cutter walks around the vehicle.

There is a large, slightly torn, Al Gore for President bumper sticker on the rear bumper.

Kitty starts the car up, it blows out orange smoke and a bunch of french fries.

                    KITTY
          I did a little work on our car.
          Hope you like it.

Cutter steps back.

                    CUTTER
          Sure.

The black sedan is parked and waiting with two MPs.

                    KITTY
          I'll try not to lose them.

                    MP
          Thank you, Ma'am I appreciate
          that. What you running on in
          there? Sure smells good. Like
          something my grandma used to
          make.

                    KITTY
          This baby burns everything from
          french fries to bacon grease.
          It's all biodegradable. Proud to
          drive one.

                    MP
          That right...Miss. Alright.

Kitty drives away, SCREECHING the tires.

INT. MILITARY POLICE VEHICLE - DAY

                    CUTTER
          Let's go...follow the smell of
          french fries.

                    MP
          Yes, sir.

EXT. KITTY'S APARTMENT BUILDING - DAY

They race through the streets until they pull up to Kitty's
apartment.

A six story apartment complex. She lives on the top floor.
Kitty gets out of the car with stacks of papers and a bunch
of large rolled up maps under her arm. There is one rose
rolled in the center.

>   KITTY
>   I'll run up...wait.

>   CUTTER
>   Come up?

Kitty hands the maps to Cutter, then continues up...

>   KITTY
>   Take these.

>   CUTTER
>   What?

>   KITTY
>   I'll be right back--don't give me
>   that look. I've waited three
>   months. Just going to have to
>   wait.

Kitty exits quickly.

>   CUTTER
>   Just...

The MPs laugh, light cigarettes and wait. Cutter's phone
RINGS.

>   Cutter (cont'd)
>   (on phone)
>   Yes, General...right away.

Cutter walks toward the front door.

INT. KITTY'S APARTMENT - DAY

Kitty enters her apartment. The walls are lined with books
and world maps from different eras. She is immediately
greeted by her small BARKING dog, MISTER, a ten month old
miniature YORKSHIRE TERRIER PUPPY. She holds the dog up and
KISSES her. Lots of licks. Then...

>   KITTY
>   Mister.

The phone BEEPS, Kitty answers.

>   KITTY (cont'd)
>   (on phone)
>   Hello? Mitsy, I was just going to
>   call you. You won't believe what
>   happened today. Big long shot.

Kitty talks while packing.

The camera follows KITTY from behind.

                    KITTY (cont'd)
               (on phone)
      Guess.

Kitty throws a little squeaky red toy key, which Mister runs
after and usually retrieves quickly. But Mister doesn't
always want to give up the toy. They play a little tug of
war.

                    KITTY (cont'd)
               (on phone)
      Ha. No! John showed up. That's
      right. Good girl.

Kitty runs around--packing her bag. We see her making
decisions about which outfits she likes in the mirror. Yes or
no. She quickly piles the clothes into a black bag.

                    KITTY (cont'd)
               (on phone)
      I'm going with him on a secret
      mission. I think. You do? Yes.
      Sure.

We hear Cutter WALKING up the steps.

Kitty rushes and puts Mister into the carrying case.

                    KITTY (cont'd)
               (on phone)
      That's right, a big Submarine.

Cutter walks up the steps, talking out loud.

                    CUTTER
      Kitty...come on, we don't want to
      keep the Navy waiting....unless
      of course you've changed your
      mind.

                    KITTY
               (on phone)
      I'll be right out.

Cutter KNOCKS and opens the door and walks into the
apartment.

                    CUTTER
      The place seems different.

Cutter looks puzzled by the sound of a small dog barking off
screen. Cutter walks toward the bedroom. Kitty cuts him off.

                    KITTY
               (on phone)
      Mitsy, I've got to go.

                         CUTTER
          How's Mitsy?

                         KITTY
          Fine. She asks about you.

                         CUTTER
          Right.

Cutter reaches for the bag containing Mister. Kitty grabs it.

                         KITTY
          I've got this one.

As they walk out we can see Mister's little nose in one of
the holes.

INT. MILITARY POLICE VEHICLE - DAY

The Navy vehicle is still running. Kitty and Cutter enter.

                         MP
          Okay. Ready.

MP on cell. Puts it down. SNEEZES.

                         CUTTER
          Listen...I'm not sure of this
          mission. Maybe you should just
          bow out.

Cutter turns to Kitty.

                         KITTY
          You men are so predictable.

MP SNEEZES again.

                         CUTTER
          This is no picnic. I'd feel much
          better if you just stayed--you
          know--don't go.

Cutter fixes her label hanging out. They have a moment of
connectedness.

Then Kitty slowly starts to react.

                         KITTY
          Better? Really? Why better?

Kitty adjusts his tie. Like she could strangle him.

                         KITTY (cont'd)
          Safe...you think I like it safe
          (she gets in his face)I like it
          dangerous.(beat) Real dangerous.
          Danger is my middle name. You're
          lucky I'm even talking to you.
          I'm not letting you get away that
          easy--I'll show you danger. Don't
          show up for my engagement party.

                              MP
          Excuse me. Sorry, that's odd. You
          ain't got a dog back there?

When the MP says...dog. Mister BARKS and sticks his head out
of the travelling bag.

                              MP (cont'd)
          I knew it.

                              CUTTER
          A dog? Just as I was
          saying...glad to um...um... have
          you aboard. This is going to be
          some mission. Some really great
          mission. You can't bring a dog.

                              KITTY
          I couldn't leave her. She's only
          a puppy.

Kitty holds Mister up to Cutter. Mister barks and tries to
bite. The MP opens the window. Puts Mister back in the bag.

                              KITTY (cont'd)
          She's very sensitive.

Kitty gets in closer and wraps her arms around Cutter. Kitty
and Cutter slowly start to kiss.

                              KITTY (cont'd)
          I love this uniform (Kitty plays
          with his metals)...Captain John
          Cutter (beat) and a secret
          dangerous mission with you
          sounds...

Cutter pushes her away.

                              CUTTER
          ...We're not James Bond and Mata
          Hari--now behave.

                              KITTY
          We'll be locked up tight in that
          sub...close. Just the three of
          us.

                              CUTTER
          Three?

                              KITTY
          Mister.

                              CUTTER
          And I thought you didn't like
          subs?

Mister BARKS.

                              CUTTER (cont'd)
          Driver, where are we going?

                         MP
          Sir, sorry, that's classified.

                         CUTTER
          Really?

                         MP
          But don't worry---I know where it
          is.

                         KITTY
          Makes me feel proud.

                         MP
          Captain, this is for you.

The MP hands Cutter a large grey envelope stamped, TOP
SECRET.

                         KITTY
          What has this got to do with
          Global warming?

                         CUTTER
          You'll see.

The car stops. We are at the port, in front of a large
building.

                         MP
          We're here.

EXT. MAUI - COAST HIGHWAY - DAY

Trees, birds and beautiful flowers line the road. Loud rock
music in the background.

The POPALISKY family driving in a rented JEEP. This is their
first ever family vacation. From New Jersey.

TOM, 30s, the driver, father, laid back, rather be watching
sports on TV than anything. Checking scores on a racing form
while driving.

MARY JEAN, 30s, the loving mother, serious and on vacation
only mode. Circling items on the map.

SOFIA, 16, cute, the teenage girl, nervous, sweet, shy.
Texting friends.

WILSON, 12, curious, the explorer, friendly. Taking notes.

WILSON points to NEPTUNE'S GIFT SHOP on the beach.

EXT. MAUI - NEPTUNE'S GIFT SHOP - DAY

The store is covered with NETS, SEA HORSES, KING NEPTUNE
STATUES, MERMAIDS, STAR FISHES, TAXIDERMIC SWORDFISH and
other OCEAN LIFE.

There is a large billboard of a HUMPBACK WHALE breaching and a BABY CALVE in the front yard.

                    WILSON
          Whale watching...can we stop? You
          said we could go on a whale
          watching tour. It's Humpback
          season...see. Guaranteed. It's
          educational.

                    MARY
          Well...Tom?

                    TOM
          I wonder if they have satellite
          TV.

                    MARY
          Tom, you promised.

                    TOM
          Come on give me a break.

There are two dogs making wild HOWLING sounds right by the front door.

INT. MAUI - NEPTUNE'S GIFT SHOP - DAY

THE CASHIER, 17, greets them as they enter.

Inside the gift shop we see a sign that says "GREEK MYTHOLOGY TOLD THROUGH SHELLS."

We see shell sculptures labelled: HERCULES, POSEIDON, ULYSSES, PYTHAGOREAS, ZEUS ON OLYMPUS.

The camera pans across a large diorama.

The diorama has a title plate which says "SKYLLA(SILL-uh) & CHARYBDIS(kah-RIB-dis)."

It is a very large and vicious looking six headed twelve legged creature with sharp teeth grabbing a sailor out of an ancient boat made of shells. There is a small whirlpool in the background.

Wilson picks the diorama up and looks it over. He accidently cuts his finger on a sharp tooth, drawing blood. He quickly puts it DOWN.

He then picks up a man made of shells on a skateboard and shows it to his mother, she LAUGHS and walks to the counter.

The dogs start BARKING.

                    WILSON
          Mom, the whales?

                    MARY
          Honey...

                    TOM
          Well?

                    MARY
          Sofia, would you like to see the
          whales?

                    SOFIA
          What? Go out on a boat? Are you
          crazy? You guys are so weird. I
          swear.

Sofia is busy TEXT MESSAGING and sending photos her friends.

                    MARY
          Will you please stop swearing and
          who are you texting? Why do I
          even ask? Where do we sign up?

Mary walks over to the cashier

                    THE CASHIER
          Pay here. I'll let them know.
          Which boat do you want? The
          Whalefinder or Betty's Congas?

                    MARY
          Tom?

Picks up a set of bongos sitting on a shelf. Taps them
lightly.

                    TOM
          Well, I am the drummer of a
          former rock band, Tom Popalisky,
          percussion. I'm sure you've heard
          of us...we had that hit "Nuclear
          Winter - Summer, Summer, Summer."
          Alright, Betty's Congas. I love
          the congas. Who's Betty?

                    THE CASHIER
          Betty left Martin cause he played
          the congas.

                    TOM
          Oh.

                    MARY
          Smart woman.

The cashier smiles. Tom puts down the congas.

                    WILSON
          Calves?

                    THE CASHIER
          That's right. It's just down the
          road at the dock. Have fun.

The cashier hands them four tickets.

Then everything begins to shake. Loud THUNDERING sounds. Dogs
BARK.

                    MARY
          What's happening? What's going
          on? Is that a subway? Hold on.
          Children come here.

Wilson and Sofia run to their mother.

                    THE CASHIER
          We're having an earthquake! Grab
          something. Stand in the doorway.
          Quick everyone. (She starts to
          pray in Hawaiian)

Mary stands in a fake Roman arch.

                    MARY
          Earthquake? My first vacation in
          eleven years...I should have
          known it. I told you. They warned
          me about Hawaii. But I wouldn't
          listen. I never listen. My whole
          life. My mother warned me.

                    TOM
          For Christ sakes.

Merchandise falls off the shelf. Lights sway.

                    THE CASHIER
          They usually don't last this
          long.

INT. NOAA - CONTROL CENTER - DAY

LIVE from inside the National Oceanic Atmospheric
Association, NOAA, Control Center, Washington D.C. The main
facility for monitoring weather systems.

DR. FRANK FARNSWORTH, Climatologist, 45, expert on global
warming, leads the group, bachelor, nerd.

HARRINGTON, 20s, great assistant, very attentive. She is
secretly in love with Doctor Farnsworth.

                    HARRINGTON
          Doctor Farnsworth, we're picking
          up another strong earthquake.
          Look at the readout. Honey Bear.

The machines are PRINTING.

                    FARNSWORTH
          Harrington!

On the screen we see EARTHQUAKE WARNING.

The room starts to shake...lights go on and off. On the wall
is a large flatscreen which reads...

Richter scale in large numbers 7.1, 7.2, 7.3, 7.5, 8.0.

> FARNSWORTH (cont'd)
> It's going to be an eighter. At
> least eight point two.

The numbers continue to rise...EIGHT POINT FOUR -- Everyone
holds on. The lights are swaying.

> HARRINGTON
> It's the end of the world...and
> we're together. Together just
> like my dream. Oh, oh, oh it's a
> premonition.

> FARNSWORTH
> Calm down, it stopped...Quick,
> get coastal warning.

All of a sudden there is an enormous aftershock making a loud
THUD.

> HARRINGTON (ON PHONE)
> Oh, Doctor. I'm so glad you're
> here. Coastal warning.

The numbers on the SCREEN rise up quickly to:

> HARRINGTON (cont'd)
> Doctor. Ten point two!

> FARNSWORTH
> Can't be.

All the phones in the office start to RING. A flashing sign
reads TSUNAMI WARNING. Bells and whistles go off.

INT. MAUI - NEPTUNE'S GIFT SHOP - DAY

> MARY
> It stopped. Thank goodness(she continues
> praying mumbling).

A second later we hear another LARGE CRASHING SOUND, it is a
very singular and sudden THUD. The aftershock. Big and loud.

> MARY(cont'd)
> It's starting again. Wilson, be
> careful. This is no way to start
> a vacation. I can not believe
> this.

Close-up of the diorama of Skylla and Charybdis which reveals
that it is still intact. There is blood on Skylla's teeth.

The dogs BARK.

Wilson walks over and pets the dogs.

> THE CASHIER
> They are going to get some pretty
> large waves from all of this.

The dogs QUIET.

The store is a mess. The damage from the earthquake is everywhere. The cashier starts to clean. A large statue of Neptune has broken in half, shells are everywhere. The statue of Ulysses is fine, a spotlight illuminates his crown of shells.

EXT. MAUI BEACH FRONT - DAY

Coast of Maui, giant waves.

ROSCO, 30, RETIRED STOCKBROKER, Out of the boiler room and into the surf. Super pro surfer.

JENNY GERONIMO, 23, HAWAIIAN, long black hair, shells around her neck, happy and one with the ocean. Best female surfer in the world.

Jenny and Rosco are adventurous. They push out further to an enormous wave.

                    JENNY
          Rosco...Check the surf...see.
          told you there would be some
          giants. Ready for some monsters.

                    ROSCO
          You are the best female surfer in
          the world I hear. This sure beats
          working on Wall Street. Trading
          waves. Look at that. I'm never
          going back.

                    JENNY
          Do you feel it?

                    ROSCO
          I feel it. I feel it. Look at
          those monsters.

Large waves roll pass them.

                    JENNY
          Strange...usually they come from
          that direction.

                    ROSCO
          That's right.

EXT. MAUI - BOAT LOADING DOCK - DAY

A long wooden dock filled with colorful locals and tourists.

Two WHALE SIGHT SEEING BOATS are preparing to go out. They are loading gear and food items.

The Popaliskys walk onboard Betty's Congas.

Betty's Congas have two very large--fifteen foot fiberglass CONGAS suspended from the HIGH DECK area on either side.

Wilson is wearing a pair of binoculars and holding a Global
Position Satellite device(GPS).

INT. WHALEFINDER - DAY

MISS TAMMY MCFADDEN, Captain of the Whalefinder, 50, life of
the party. Passing out Whalefinder T-shirts in an exuberant
manner.

There are a dozen tourists on board taking photos, talking,
and snacking.

                    TAMMY
              (on the 2-way radio)
          Marty, That was some shaker.
          We're going to head east a
          little. No earthquakes on the
          ocean. Although.

Flocks of birds are heading west.

INT. BETTY'S CONGAS - DAY

MARTIN "MARTY" MONSANTO, Captain of Betty's Congas, 50s
from New Jersey, out of shape. Already a little drunk. Father
of Lucas.

LUCAS MONSANTO, 17, fascinated by whales, girls, surfing.
Tall and thin.

Martin is watching the birds. Shaking his head.

                    MARTIN
              (on the 2-way radio)
          Tammy, I'm thinking west. Over.

INT. WHALEFINDER - DAY

The boat is full of tourists shaking Tammy's hand and taking
photos.

                    TAMMY
              (on the 2-way radio)
          Well, Marty, something tells me
          I'm going to be especially lucky.
          How 'bout dinner with drinks says
          I see the first whale of the day.
          Over.

INT. BETTY'S CONGAS - DAY

                    MARTIN
              (on the 2-way radio)
          You're on. But let's make it a
          little more interesting...say it
          has to be a mother and a calve.
          Over. Lucas. Must be a confirmed
          sighting. Going to be making some
          of that Bar B Q Mahi Mahi, I
          hope.

INT. WHALEFINDER - DAY

                    TAMMY
              (on the 2-way radio)
          Deal. Over and out.

A tourist holding a cracker, walks up to Tammy.

                    TOURIST #1
          What's this?

Tammy tastes it.

                    TAMMY
          Hummus.

Tammy pushes the speed control all the way down, the boat
takes off. The tourists fall back into their cushions.

INT. BETTY'S CONGAS - REAR DECK - DAY

The Popaliskys gather around.

                    MARTIN
          Good morning. I am your Captain
          Marty Monsanto and this is my
          first mate and son, Lucas.
          Welcome Popaliskys. Lucas say
          hello. Don't scare them.

                    LUCAS
          Good morning everyone. Welcome
          aboard.

Lucas tips his hat to Sofia. Now---she is interested.

                    MARTIN
          Sit back...enjoy the ocean air.
          We should be running into some
          humpbacks soon. I don't mean
          we're really going to run into
          them. It's their season to return
          here. We'll be seeing breaching
          and frolicking and I'm sure
          plenty of baby calves. Lucas...
          tell them about the whale cam.

Lucas and Sofia's eyes lock in some special kind of soul mate
way. Everything Lucas says is now directed to Sofia. He is
newly enthused at his job. Martin is drinking.

                    LUCAS
          We got a whale cam that lets you
          see underwater, (Softly to Sofia)
          Would you like to try?

                    SOFIA
          Sure.

INT. BETTY'S CONGAS - BELOW DECK - DAY

Lucas and Sofia go below deck into a large section of the
boat. There are orange cushions to sit on.

                    LUCAS
          Careful, the steps are steep. I'm
          Lucas.

                    SOFIA
          I heard. Sofia.

                    LUCAS
          A beautiful name!

Lucas helps Sofia down the steps.

                    SOFIA
          You're not from here?

                    LUCAS
          Jersey. But now...relocated to
          Maui, my Dad's idea of paradise.
          Daiquiri's till he drops. Here,
          this is the whale cam.

Sofia looks over the whale cam, a black metal pair of
binoculars attached to an underwater video system.

                    SOFIA
          We're from Newark. You fish?

Sofia picks up a small net with a big hole in it.

                    LUCAS
          I'm more into swimming with the
          whales.

There is an faded Whale poster on the wall.

                    SOFIA
          Really? I've wanted to do that.
          What's it like? Being that close
          to such large animals.

Lucas gets closer to Sofia.

                    LUCAS
          Sometimes, they swim real
          near...and look right at you with
          those big eyes. Big beautiful
          eyes. Smart eyes, Like yours. (He
          gets in closer) Here look at this-
          --there are some humpbacks
          swimming in the far distance.
          They have long flippers like
          arms.

They get closer. He's getting ready to kiss her...when from
above...

                    MARY
          Sofia...Come back up here!

Upper deck: Tom to Martin.

                    TOM
          Nothing quite as loud as a mother
          YELLING.

                    SOFIA
              (apologetically)
          I have to go.

Sofia exits...(beat)

Martin HOLLERS from above.

                    MARTIN
          Lucas? Get back up here. Check
          that everyone is wearing their
          life preservers. Leave the
          tourists alone.

Sofia catches Lucas out of the corner of her eye, smiles.

INT. WHALEFINDER - DAY

Captain Tammy lets a tourist take the wheel while she takes a
photo of her. As she focuses the camera Tammy begins to
notice large waves and the rocking of the boat.

                    TAMMY
          Hold on. Getting thick like soup.

There is a heavy fog rolling in. Tammy checks her compass
which goes back and forth between on and off.

                    TAMMY (cont'd)
              (on the 2-way radio)
          Coast Guard? Hello? Over.
          (Static)
              (Gets cut off)

The Whalefinder is at the edge of the Maelstrom Whirlpool.
Giant waves surround the boat. People are screaming. Strange
fields lifting water vapor in circular motion.. Everything
electrical goes DEAD.

We follow the boat through a dark cone of downwardly
spiralling ocean.

                    TAMMY (cont'd)
          What? What is happening? The
          ocean is...(grabs the radio) May
          day. May....Hello? Martin?
          Martin? Mayday. Marty..I love
          you..

The ROARING SOUNDS of the CRUSHING WATERS are deafening.

The Whalefinder gets pulled into the Whirlpool.

INT. NOAA - CONTROL CENTER - DAY

                    HARRINGTON
          I have a reading of a cyclonic
          disturbance near that emergency
          call. It's off the coast of Maui.
          It's present location is one
          hundred fifty four point one two
          five west and nineteen point six
          zero north.

The Whirlpool appears on the large screen.

                    FARNSWORTH
          I'll be right back.

                    HARRINGTON
          Wait.

Harrington hands him a cell phone.

                    FARNSWORTH
          Report all that in. I'll be back.
          Don't worry.

Harrington types, 154.125 West and 19.60 North.

                    HARRINGTON
          There is Coast Guard in that
          area. I'll put a call in.

Farnsworth exits.

INT. COAST GUARD VESSEL - LATER

COAST GUARD CAPTAIN ALAN, 50, tough, short, Napoleon like,
With binoculars in hand.

A large 110 FOOT COAST GUARD BOAT off the coast of Maui.

                    ALAN
          We're right in the area where
          that distress call just came in.
          Anything? Anyone?

Captain Alan sees edge of the Maelstrom.

                    ALAN (cont'd)
          What the ...Fast reverse, reverse
          all engines to maximum.

The giant circular current is filled with churning white foam
and a heavy ROAR. Gargantuan waves and ribs of water.

The ship goes around and around.

                    ALAN (cont'd)
          More power.

The ROAR of the Whirlpool masks the sounds of the engines at
maximum. The vertically spiralling waves are bellows of white
foam.

                         ALAN (cont'd)
              I can't hear anything. Reverse.
              Reverse full. Fast.

INT. RESCUE HELICOPTER - OFF MAUI - DAY

A Coast Guard Rescue Helicopter follows. CAP, the pilot, 40s,
and DALLAS, 30s, the copilot both wearing helmets.

                         CAP
                   (into the headset)
              This is--Coast Guard Rescue.
              Captain Alan it looks like you
              are almost clear. Look at the
              size of that thing.

INT. COAST GUARD VESSEL - DAY

Captain Alan looking at the Whirlpool through his binoculars.
On the phone. Holding on.

                         ALAN
              We're breaking free. Relay this
              to NOAA.

INT. NOAA - LECTURE HALL - LATER

Lecture hall. The room is full of various ARMY, NAVY and
MARINE personnel, including:

NAVY GENERAL GEORGE T. CHERRY, 50s, stern, tight fisted and
with authority. Perfectly dressed and pressed but suit too
tight. High blood pressure.

Two Seals, BRICK and MANELLO, 30s, hardened and ready to
start a fight. Dangerous.

KITTY enters.

                         CHERRY
              Let me introduce Doctor
              Katharine, I mean Kitty Honeywell
              from the Navy's Department of
              Oceanography. The leading expert
              on ocean level and climate
              change. Doctor?

Kitty stands behind podium. There is video with the
following.

                         KITTY
              Thank you. Right now this
              Maelstrom Whirlpool is thirty or
              forty miles across and growing.
              It is the largest whirlpool ever
              recorded. A hundred times larger
              then the famous Moskstraumen off
              the Loften islands in Norway.
              Bigger than the vortex that
              swallowed Lake Peigneur in
              Louisiana after the Texaco
              drilling accident.

                    KITTY (cont'd)
          And bigger than any glory hole
          I've ever seen from a Dam
          emptying.

Kitty walks over to Cherry and examines the photographs he
has in his hands. She grabs a large plastic triangle placing
it over the photograph

                    KITTY (cont'd)
          And it's on a collision course
          with the Hawaiian Islands. Headed
          right towards the main population
          center. Potentially killing more
          than 170,000 people that live on
          the coast. Seventy two hours
          before it hits. General?

Manello and Brick are off to one side talking.

                    MANELLO
          These Navy guys have it made.

                    BRICK
          Sure do.

Just as Cherry is about to continue Cutter runs in.

                    CUTTER
          Sorry to be late.

                    CHERRY
          And of course Captain
          Cutter...who I believe you all
          know. The expert of submarine
          testing. Winner of three Navy
          crosses and a silver star. Will
          be in charge of the mission. But
          we'll get to that. Oh. One more
          thing--at the rate it is growing
          it will be on the Pacific fleet
          in about the same three days.
          Farnsworth?

A stillness fills the air.

                    FARNSWORTH
          At first we thought that this
          would just fix itself but it now
          appears that it is being
          generated from a hole deep under
          the crust. We estimate this is
          causing the vacuum which is
          pulling in water at an
          astonishing rate of 45 million
          pounds of vacuum in a radius of
          just a few feet. And every once
          in a while it stops and then
          starts up again.

                    CUTTER
          Kitty, why do you think it's
          happening there?

                    KITTY
          Major fracture zone. See the
          model.

We see this demonstrated in a holographic model of the sea
floor bed.

                    FARNSWORTH
          Yes. We also thought that this
          might be the planets way of
          adjusting for the increase in
          water height due to the melting
          of the polar ice caps from global
          warming.

We see this in another holographic model of the ocean levels
and coast lines.

                    KITTY
          By what? Sucking it into the
          interior of the Earth? What are
          you talking about?

The SEALs are delighted at her feisty demeanor and high five
each other.

                    BRICK
               (To Manello)
          Like to hear more about this.

Kitty gives them a look. Brick salutes.

                    FARNSWORTH
          Perhaps there is another layer of
          water in the interior.

                    KITTY
          It might just be the biggest
          change ever in the shape and size
          of the oceans. Or it might be the
          end of all life on planet Earth.
          Alright. What EXACTLY do you want
          to try?

                    CHERRY
          Fix the hole. Put a cork in it.

                    KITTY
          Right.

                    CHERRY
          Let me show you what you're up
          against.

We hear a KNOCK on the door.

The door opens and KNOCKING over a garbage can..

Enter --

FRANK DANISH, 30s, FEMA(Federal Emergency Management Agency, ASSISTANT, CHIEF OF STAFF). A little confused, slow and out of shape.

> DANISH
> Frank Danish, Excuse me...I hope I haven't missed any of the briefing. Nice to meet you all. Frank Danish, Assistant Chief of Staff...Federal Emergency Management Agency, FEMA. I have some very important paperwork I need you officers to sign off on. I'm so glad I found you all together.

Waves to everyone.

> CHERRY
> Frank, good of you to join us. We just got started. Hold off on the paperwork for a minute.

> DANISH
> Do you have any coffee? I could use a mochacchino. Just flew in from Florida. I'm so jet lagged. Oh...How's everyone doin? Macadamian?

Opens a can of nuts, spills all over the floor. No one answers.

> FARNSWORTH
> Harrington, please open the lab.

> HARRINGTON
> Yes, Doctor. It would be a real pleasure to do so. I like that tie.

Harrington presses a few buttons.

INT. NOAA - WATER LABORATORY - DAY

The center wall OPENS to reveal a brightly lit room filled with two very large glass CYLINDRICAL TANKS, one is full of water BUBBLING, the other empty.

And introducing COLONEL FENSTER, female, 50s, CIA, wearing a ruffled suit. Dark glasses. Standing off to the side, trying not to be noticed, reading through documents. Doesn't smile. Nervous. Stiff.

> FARNSWORTH
> This is how it works. Who are you?

Fenster shows her ID card.

                    FENSTER
          I've been waiting for you, I'm
          with intelligence. Under cover
          intelligence.

                    FARNSWORTH
          Never mind. Right.

Farnsworth goes over to a computer and starts to TYPE,
simultaneously, pumps and other machines start to make a
multitude of NOISES.

Inside the tank, a WHIRLPOOL starts to form.

                    FARNSWORTH (cont'd)
          Watch. This is a standard flow.
          This is a rapid flow and this is
          extreme flow.

As he explains this we see a model submarine go by the path
of the small whirlpool in the tank. The model is torn into
pieces.

                    CUTTER
          I'm not sure--but that doesn't
          look too good.

                    DANISH
          Who sank my battleship?

No one laughs. But they all stare.

                    CHERRY
          We're thinking that it might be
          possible to approach it from
          underneath and seal it using
          liquid Nitrogen. This hasn't been
          fully cleared yet for release.
          But it's on board now so let's
          talk about it. About six years
          ago the Office of Naval
          Operations at the CIA asked us to
          construct a vessel that could go
          through anything from ice to
          liquid magma to solid rock, sort
          of like a super armadillo.

The computer monitors show various images of the evolutions
and construction of different rescue subs.

                    CHERRY (cont'd)
          Theoretically it might be used as
          a recovery vessel for co-patriots
          or stranded miners. We made use
          of a new class of nuclear attack
          submarines. We kept it quiet
          cause it was so damn expensive.
          Instead of the government
          building four more attack subs at
          two billion each they built one
          very special one.

                    CHERRY (cont'd)
          It's what known as a RTPNACDSRV,
          a RETICULATED TRI POWERED NUCLEAR
          ATTACK CLASS DEEP SUBMERSIBLE
          RESCUE VEHICLE. We call it the
          USS LEVIATHAN. Named after the
          legendary sea monster. Cause
          it's...a monster.

Farnsworth TYPES into a keyboard and a super detailed
holographic model of the USS LEVIATHAN appears on the large
flat screens.

                    DANISH
          It cost how much?

Danish taking notes.

                    CHERRY
          It can be anywhere on the planet
          in less than two days and can go
          thousands of feet below the
          crust, deep into the magma. It is
          the only sub in the world with
          three reactors.

                    KITTY
          Three?

                    DANISH
          One wasn't enough?

                    CHERRY
          The reactors are small and much
          more efficient than the older
          models. And fully self contained.
          Solid walls.

Holographic cross sectional schematics of the USS Leviathan
nuclear reactors appear on the flat screens.

                    KITTY
          Nuclear power is awful, I could
          run this on compost and vegetable
          oil if someone just gave me a
          chance.

                    CUTTER
          I'd like to see that, a submarine
          that farts.

                    BRICK
          Where do we come in?

Brick steps forward.

                    MANELLO
          Easy...

                    FARNSWORTH
        We're thinking that we might try
        to go underneath and patch the
        hole...it's like what they do in
        some villages...when the muds
        rise from sink holes. Only we
        will use special torpedoes filled
        with liquid Nitrogen.

                    KITTY
        Fix the hole like a sink hole?

                    CHERRY
        Exactly. Put a cork in it.

Two Navy ensigns wheel out a TEN FOOT TORPEDO.

                    CHERRY (cont'd)
        These are the special laser
        guided Nitrogen torpedoes. Here.

Kitty starts LAUGHING and pointing at the torpedoes.

                    KITTY
        You think you're going to stop
        that Whirlpool with this? You
        know, size does matter.

As we see the torpedo from different angles it disappears.

                    CHERRY
        These torpedoes bore through the
        earth's crust at Mach 4...about
        twenty-five hundred knots with
        extreme accuracy and stealth
        capabilities. You're getting
        twelve of these and six regular
        attack torpedoes.

Cherry holds up a large piece of carbon carbon, even though
its immensely large, it weighs practically nothing.

                    CUTTER
        Attack torpedoes?

                    CHERRY
        Never can tell.

                    DANISH
        Do we know the cost of these?

Farnsworth hands Danish a slip of paper, it is the actual
bill of sale from US Boat.

                    FARNSWORTH
        Here. The torpedoes bore through
        earth, ice or rock.

Danish looks closely.

                         DANISH
          Eight billion three hundred forty
          seven million two hundred nine
          thousand dollars and twenty one
          cents, in adjusted dollars.
          ($8,347,209,000.21)

We see a video of four different torpedoes demonstrating just
how much faster this torpedo is than the others.

                         CUTTER
          I see.

                         KITTY
          Boring through the crust does not
          sound safe to me. What about
          damage to the sea floor?

                         CUTTER
          It'll seal itself.

                         KITTY
          Really.

                         CUTTER
          Well.

                         CHERRY
          The torpedoes follow a tunnel
          created by a massive carbon
          dioxide pulsed laser.

On the viewing screen we see a simulated composite imaging of
a laser created path and the torpedo going through it.

On another screen we see a three dimensional imaging of the
USS Leviathan going through rock at a very fast speed.

                         KITTY
          Any material?

                         CUTTER
          Five times the speed of the last
          submarine I tested. Like an
          underwater Ferrari.

                         CHERRY
          The spinning gears create a
          tunnel of low pressure through
          which the sub glides almost
          friction free.

We see a flash of the image of the ship's many gears.

                         CHERRY (cont'd)
          The first prototype moved in
          speeds in excess of 150
          knots...over 100 mph.

We see an image of the prototype moving quickly through the
ocean.

                    CUTTER
          What happened to that one?

A man who has been hiding in the doorway steps forward. Chief
Petty Officer, IAN O'BRIEN, 50, tattooed, from Belfast, he is
as tough as he appears. Polished and clean shaven, moves
quickly.

                    O'BRIEN
          Chief Petty Officer Ian O'Brien
          here. We got hit pretty bad,
          squashed like an accordion. Ship
          couldn't hold the pressure.

                    CHERRY
          She was buried underground. Half
          of the men made it out alive.

                    O'BRIEN
          Not exactly half and not all of
          us made it to the escape pod. I
          lost some good friends. We
          climbed and pushed our way
          through the twisted metal.

O'Brien rolls up his sleeves showing scars.

                    CUTTER
          Damn DSRVs. Give me an attack sub
          and some shallow water and I'll
          show you damage.

Cherry walks towards Cutter.

                    CHERRY
          Go underneath the Whirlpool...
          Find the hole and as you said
          before-plug it.

                    FARNSWORTH
          I need to warn you. Even if we
          seal the hole we're not sure what
          might happen...it might keep
          reappearing. It just might split
          into smaller whirlpools. Just see
          what happened to these tornadoes.

He demonstrates this on the computer screen above. We see
tornadoes splitting and separating into smaller ones and
those into even smaller ones.

                    CHERRY
          That's where the SEALs come in.
          There is one nuclear torpedo
          onboard as well. You will need a
          special key to operate. Here.

Hands Fenster the red key.

Video of large underwater nuclear BLAST. BLOWING everything
out and pulling it back in.

Kitty takes out the pointer.

                    KITTY
          Are you kidding? You must be
          spending too much time underwater
          cause that is a ridiculous idea.
          A nuclear blast is going to cause
          an even bigger mess. The
          Whirlpool happens to be located
          near the Molokai ridge, a major
          fracture zone. We'll trigger
          another even stronger earthquake
          if we're not careful.

                    DANISH
          Not another earthquake.

Danish drops all his papers on the floor.

                    DANISH (cont'd)
          Shoot.

                    FARNSWORTH
          She's right a blast near the
          ridge might be catastrophic.

                    CHERRY
          We're thinking that the Nitrogen
          might freeze enough magma to stop
          the suction. The magma then seals
          the hole. End of story.

On the monitor we see this computer simulation of a cross-
section of the earth.

                    KITTY
          Anything else?

                    CHERRY
          There is only one other catch.
          The Nitrogen is volatile and the
          torpedoes must be loaded twenty-
          three minutes before you want
          them to detonate. Otherwise they
          will fracture the tube and
          explode. Cutter?

Monitors play this explosion back and forth. One from inside
the sub where the tube fractures and blows up the submarine
and one from outside showing the pieces of the sub scattering
in all directions.

                    CUTTER
          Underground?

Cherry shakes his head.

                    CHERRY
          You'll be accompanied by Colonel
          Fenster and her SEAL team.

                    FENSTER
          Right.

                    CHERRY
          Kitty, you too. In case
          Oceanographic information is
          needed.

                    KITTY
          I figured.

                    CHERRY
          Danish?

                    DANISH
          I can't...I get very seasick.

                    CHERRY
          You will be taken to the sub
          immediately.

INT. NOAA - WATER LABORATORY - DAY

The General stands up and then the rest of the group rises.

                    CHERRY
          Let's go.

They all begin to exit.

                    BRICK
          Looking forward to seeing you on
          board ma'am.

The SEALs leer at her.

                    CUTTER
          Kitty.

                    KITTY
          John...don't worry. I'm sure you
          can take both of them.

She grabs his BICEP.

                    KITTY (cont'd)
          Solid. I like that. What's this?

They turn and see the TV which having an EMERGENCY NEWS
BROADCAST.

INT. NEWSROOM - MAUI - DAY

TV NEWS REPORT over Rosco and Jenny surfing.

                    REPORTER
          This is a special bulletin from
          action news room. There is a
          report of a giant whirlpool off
          the coast of Maui. So far at
          least one boat is missing.

                    REPORTER (cont'd)
The Coast Guard has issued a
warning to all Mariners. It is
located near the annual migratory
route of several different
species of whales including the
humpback, Scientists believe this
might be the result of the deep
underground earthquake from
earlier today which measured as
high as 10.2 on the Richter
scale. But other scientists claim
that this is some new process in
the global warming cycle. This is
your action news room reporting.
The population is reminded to
stay away from beaches till
further notice. Especially you
surfers. You hear that?

EXT. OFF THE COAST OF MAUI - DAY

Rosco and Jenny are surrounded by large waves. Paddling
around on surfboards.

                    ROSCO
                (screaming)
I never thought I'd get the
chance to do this with you.

                    JENNY
Careful of the big ones.

                    ROSCO
I'm going to the ridge.

Rosco paddles out deeper.

                    JENNY
Wait. That's not the ridge.

Rosco disappears behind a large wall of water.

INT. MAELSTROM WHIRLPOOL - DAY

POV Rosco

Rosco has entered the outer edge of the Maelstrom Whirlpool.

                    ROSCO
I feel it.

Rosco surfs around and around spiralling downward through a
funnel of ocean.

EXT. UNDISCLOSED COVERED DOCK - NIGHT

They open the car door to reveal the empty dock. Cherry,
Kitty and Cutter step out.

                    CHERRY
Welcome to our little community
here.

                    CHERRY (cont'd)
          I'll call them (takes out his
          phone). Hello? We're here. New
          plan includes underwater minutes.

EXT. USS LEVIATHAN - NIGHT

A moment later the USS LEVIATHAN starts to rise out of the
ocean. It is comprised of a series of black carbon carbon
fiber concentric spinning blades. The entire reticulated body
spins and then slowly comes to a stop.

It is enormous next to them. At least five stories tall.

View of giant screw like front end.

View of side of sub.

A part of the gear opens up to a doorway.

The doorway reveals a dark red interior light.

A black gangway comes forward.

INT. USS LEVIATHAN - NIGHT

ROYAL MCCABE, EXEC, Executive Officer, 50s, pressed suit, by
the book, married. An old college friend.

Royal walks out--they salute each other. Standing behind them
is...

DR. EMMA KEATING, MD. 30s, Medical Doctor, from the Midwest.
Polite, modest.

                    EXEC
          Welcome aboard Captain and Miss.,
          Executive Officer, Royal McCabe.
          At your service. We've been
          waiting for you. Welcome aboard
          Doctor.

Royal steps aside.

                    KITTY
          Kitty, please.

                    CUTTER
          Good to see you Royal. How's the
          family?

                    EXEC
          John, good to see you, everyone's
          fine. Ready to disembark.

Frank extends his arm out. With his FEMA ID card.

                    DANISH
          Frank Danish, I'm with FEMA, the
          Federal Emergency Management
          Agency. I was told you had a
          cappuccino machine. I need to
          examine.

                    EXEC
          Your quarters are ready. Follow
          me. Cappuccino? I'll show you the
          cappuccino machine. Dr. Keating
          will show you to your quarters,
          Doctor Honeywell.

                    KITTY
          Kitty.

Emma Keating steps forward.

                    KEATING
          Emma Keating, Two female Doctors
          onboard, nice to meet you. This
          way, Kitty. I think we've got
          them outdone in the brain area on
          this mission.

                    KITTY
          I think you're right.

Kitty and Keating walk down another corridor laughing.

INT. USS LEVIATHAN - HALLWAY

                    CUTTER
          Let's check the control room.

                    EXEC
          Yes, Sir. Danish...Odd name. The
          cappuccino machine is down this
          hallway to the end...then take
          the staircase down two flights
          and go to the end of the corridor
          where the Mess Hall is. Not the
          service elevator. Do not take the
          service elevator. Understand?

                    DANISH
               (taking notes)
          Got it. Go down two floors to the
          cafeteria. Easy. Oh, and by the
          way, my parents owned a bakery in
          Canarsie hence the name.

                    EXEC
          Straight ahead..

Danish walks away slowly.

                    DANISH
               (outloud)
          Just down the corridor. Coffee.

Danish walks by many ELECTRONIC SIGNS AND SENSORS. Ignoring
warning lights, taking notes.

                    DANISH (cont'd)
          Oh...an elevator.

Danish steps inside the elevator.

A sexy sweet female VOICE greets him.

INT. USS LEVIATHAN - ELEVATOR - CONTINUOUS

                    SUBMARINE VOICE
          Good day, Sir. What level please?

Takes out a note pad. Reads.

                    DANISH
          Hi...two floors down, Miss..

                    SUBMARINE VOICE
          Malfunction. Storage. Elevator
          descending to basement level.
          Please hold--Emergency Release.

The elevator drops very quickly. Lights go on/off.

                    SUBMARINE VOICE (cont'd)
          Basement level. Nuclear reactor
          access. Authorized personnel
          only.

Danish walks out slowly.

                    SUBMARINE VOICE (cont'd)
          Please exit the elevator quickly.

The door SLAMS shut behind him.

And the elevator goes back up.

Danish walks out of the elevator, confused he accidently
bumps his head hard on a low bulkhead.

                    DANISH
          Oh...that's not good.

His eyes shut...he passes out.

INT. USS LEVIATHAN - CONTROL ROOM - LATER

The EXEC continues towards the rear.

They walk through the stainless steel corridors.

The ship has Nitrogen containers everywhere. They steam off
an ultra cold vapor. Many are stacked on top of one another
creating small hills of ice.

As each man is introduced they give him a high five greeting.

                    EXEC
          Over to the right is navigational
          equipment, to the right of that
          is the NAV star GPS receiver,
          here's the ships control system,
          the helmsman, and the planesman,
          and then along this wall is the
          weapons control panel.

                    EXEC (cont'd)
          The plotting tables and the TYPE
          3 periscope.

                    CUTTER
          Got it. Same as the Sacramento.
          Here's our orders.

Cutter walks over to the plotting table and hands the EXEC a
small FLASH DRIVE. He takes the flash drive and inserts it
into a small slot. This activates a host of systems on the
Leviathan.

                    EXEC
          Thank you. Ready to disembark.
          Sir?

The Control room is filled with activity. The crew work
on various computer monitors all wearing headsets with
virtual 3-D projections

General Cherry enters. Shaking hands, saying his good-
byes.

                    CHERRY
          You have your orders. Good luck men.

                    CUTTER
          Thank you, Sir.

                    CHERRY
          Cutter.

Cherry pulls Cutter in close, WHISPERS into his ear, quietly.

Then...loudly --

                    CHERRY (cont'd)
          Have a successful mission-the
          world is counting on you and come
          back safely, oh and please don't
          destroy this sub too.

Cherry exits. Then the SEALS come in.

                    FENSTER
          Men.

                    EXEC
          Roger with that. One minute to
          disembark. Disembarking. Hold on.
          Sound horns.

The HORN blows loudly and distinctly three times.

                    CUTTER
          Ten hours since the last report,
          status?

                    EXEC
          Nearly a hundred miles across and
          a mile deep. And it's also
          swaying in different directions.

Various monitors showing the different sides of the Maelstrom
Whirlpool. Stretching from one part of the ocean to another.
Layers of fog spiralling above.

INT. USS LEVIATHAN - NIGHT

                    EXEC
          I want to introduce you to your
          steward, and the Bowman
          Scholarship Program Intern,
          Robert Fish.

                    CUTTER
          Steward Robert Fish? Good. Let's
          inspect the rest of the boat.
          Frank Bowman scholar, how old are
          you?

                    FISH
          Twenty two, Sir.

                    CUTTER
          Fish?

Cutter has taken out his map and is trying to follow along.

                    FISH
          Here you go. The map goes like
          this, it's sort of upside down
          and reversed, from a helmsman
          point of view. Follow me.

As Fish walks through the doorway he nearly bumps his head.

                    FISH (cont'd)
          Sorry Sir, low bulkheads.

                    CUTTER
          I see.

                    FISH
          Yes, Sir. The crew berthing is
          right down here. We are running
          with twelve, a skeleton crew. Oh,
          excuse me. This way.

Fish steers away from Fenster and the Seals.

INT. USS LEVIATHAN - ENGINEERING - DAY

The Seals are in an office reading through manuals and charts
of the USS Leviathan.

                    FENSTER
          Make sure you go through
          everything. Is that understood?

Fenster grabs Brick by the collar.

                    BRICK
          Clear.

                    FENSTER
          I can't hear you. You need to be
          familiar with every bolt. You got
          that? Let's review weapons and
          triggering. Understand!

                    BRICK
          Yes, Sir. I mean ma'am.

Brick and Manello stare at Fenster.

                    FENSTER
               (screaming)
          Don't you two have something to
          do? Why am I still talking to
          you? How can we ever get anything
          done if you two loaf around. Get
          to work. I want choices, choices,
          Choices. This should be my boat.
          Go.

They turn, bump into each other and then exit.

                    FENSTER (cont'd)
          I have the worst cramps.

INT. USS LEVIATHAN - HALLWAY - DAY

Cutter and Fish continue walking through the narrow corridor
to the sleeping quarters.

                    CUTTER
          Fish, where have you served
          before? You look familiar.

                    FISH
          I was training on the Springfield
          when I heard a rumor about the
          USS Leviathan.

They have to walk around the Nitrogen tanks in the hallway
and grab the railing to not slip on the ice.

                    CUTTER
          Liked it so much you decided to
          stay.

                    FISH
          I was Shanghaied. I walked in and
          they wouldn't let me leave. Would
          not let me leave. They said this
          would be the first in a series of
          tests for the Bowman scholarship
          program. Careful, low ceiling.

Fish guides Cutter.

                    CUTTER
          You're kidding? Who said that?

                    FISH
          Royal. He said that based on some
          regulation that if I am in a sub
          when it's time to go to sea, the
          officer in charge can enlist me.
          Being a Bowman scholar didn't
          matter. Seems ridiculous then I
          heard that you were going to be
          the Commanding Officer and I
          thought--my lucky day. Imagine
          that, the legend. I wanted to
          call my mom in Kentucky but they
          wouldn't let me.

Cutter stares at him.

                    CUTTER
          Royal, ha. That's what I call
          some real innovation in the
          drafting process. Welcome aboard.

Cutter gives him a pat on the shoulder. They continue
walking.

                    FISH
          The crew is right up the hall.

INT. USS LEVIATHAN - HALLWAY - DAY

                    CUTTER
          I think I'm going to sit down for
          a moment. I have to make some
          notes. Take five.

                    FISH
          Yes, Sir. I'll be right back.
          I've got to file this.

Fish holds up the chart for Cutter to sign. He looks it over,
smiles and then signs.

                    CUTTER
          Sure thing.

Cutter takes out a long list of items to check on the
Leviathan. He sits down and leans back into the cushion and
slowly his eyes close and his body relaxes.

He begins to listen to the VOICES from the men in the other
room.

Along with the voices we hear the HUMMING of engines.

The hall lights are soft red.

It is a QUIET moment. Cutter is falling asleep in the warm
seat.

Slowly Cutter's eyes close.

As Cutter begins to fall asleep he hears O'Brien TALKING to a
group of men.

                    O'BRIEN
          Are you familiar with the legend of
          Skylla and Charybdis? The Whirlpool
          and the monster.

                    FANUCCI
          Let's here this one O'Brien.

                    O'BRIEN
          A long time ago...when men were
          put on boats that travelled the
          globe, there was a spot deep on
          the Mediterranean where many a
          sailor met his fate.

Begin black and white dream sequence.

EXT. SKYLLA AND CHARYBDIS DREAM SEQUENCE - DAY

Kitty as SKYLLA the MONSTER with six heads and twelve legs
located high on a cliff overlooking the inlet reaching down
to the water. CHARYBDIS is the name of the monster WHIRLPOOL,
lower down and on the side of the inlet.

The crew on the boat: Cutter, Fenster, Fanucci, Cherry,
Danish, and O'Brien.

A ancient vessel made of exotic woods.

RAIN, THUNDER AND LIGHTNING.

2000 years ago.

The Boat: The ODYSSEY

Cutter as ULYSSES.

They are dressed in historic Greek outfits with swords,
shields and knives..dreamlike though..from Cutter's mind.

Cutter has added a beard and sandals.

                    O'BRIEN
               (voiceover, slowly)
          A terrible, twelve legged
          creature named Skylla with six
          vicious heads would guard the
          narrow coast. Each head had rows
          of shark teeth. The men would
          hopelessly try to fight her
          amidst thick fog and powerful
          electrical storms.

                    CUTTER
          She lurks inside the cave--
          beware. Stay alert.

Cutter looks through a BRONZE SPYGLASS.

The Odyssey approaches the coastline.

There are groves and groves of juicy fig trees, untouched by humans for years lining the coast.

Birds, bees, and butterflys fill the area

                    O'BRIEN
                (voice-over)
      Skylla appears from nowhere.
      Jumps at you. And when you avoid
      Skylla--Charybdis, the whirlpool
      is waiting. With it's boiling
      salt water sucking and spitting
      out everything in sight. Spinning
      around and around. A dynamo,
      crushing all that enters it.

                    CUTTER
      Men look out for the terrible
      whirlpool. Watch yourselves. It
      is a one way trip. Forget the
      large delicious figs that line
      her shore. And check for Skylla,
      she sneaks up on you.

We see the groves of trees.

                    O'BRIEN
                (voice-over)
      The legend has it that the
      sailors would avoid Skylla only
      to be swallowed alive by
      Charybdis. One boat just managed
      to escape when Skylla reached
      over and grabbed three men,
      tearing off their heads in one
      bite. She chewed on them a while,
      and then spit them into the
      whirlpool leaving the men for
      eternal damnation in the
      bottomless pit they call the dark
      sea.

                    FANUCCI
      The dark sea. Rest in peace.

Colonel Fenster as the brave but foolish SOLDIER.

                    FENSTER
      We have new intelligence that she
      poses no threat. Don't worry.

Kitty as Skylla bites Fenster's head off.

                    FENSTER (cont'd)
      ...she's got me...

Skylla eats Fenster.

                    CUTTER
      Damn you Skylla. Draw your swords
      men. Quickly.

The men take out their swords. Danish steps forward to examine a blade.

> DANISH
> Are those double edged swords?
> Couldn't we use single edged--it
> would save so much money...

Skylla bites off Danish's head, his arms start flaying. She spits his head out into the whirlpool. It slowly sinks and says...

> DANISH (cont'd)
> ...paperwork. Bubble(sounds).

> CUTTER
> Hated to see him go.

Cutter lunges for Skylla. Cutter's sword is knocked loose. Skylla reaches for him...Dozens of rows of bloody and razor sharp shark teeth

The horror makes Cutter SCREAM at the top of his lungs.

> CUTTER (cont'd)
> Stop. Check please. Help. No
> Kitty we can get married. I'm
> sorry I missed the engagement
> party. I'm sorry...I'm sorry.

Cutter still caught up in the dream -- wakes up.

End black and white dream sequence

INT. USS LEVIATHAN - HALLWAY - DAY

Cutter's hands are in motion, trying to fend off the attack.

Cutter rises. Fish enters.

> FISH
> How you doing Captain? Get some
> rest?

> CUTTER
> Fine...fine...just fine, great,
> great.

He stands up, adjusts his tie.

> CUTTER (cont'd)
> Let's go.

Cutter and Fish slowly walk into the crew area.

INT. USS LEVIATHAN - CREW AREA - DAY

The crew is sitting, prepping and resting in their cots.

Cutter walks through, looking at photos in the bunk area.

                    CUTTER
          At ease gentlemen.

                    O'BRIEN
          Sorry, Sir. We were just telling
          some old sea tales.

                    CUTTER
          Yes, I heard. Let's hope that's
          the case. Don't want to lose my
          head.

                    O'BRIEN
          I'd say...Lose your head? You saw
          her? The man-eating she-beast
          with six heads.

O'Brien shines his shoes, his eyes transfixed on some distant
object.

                    CUTTER
          Who?

                    O'BRIEN
          You know. Skylla, the beast.

Cutter freezes.

                    CUTTER
          It's nothing.

                    O'BRIEN
          Did she get you?

Cutter hesitates to answer. Looks over some papers.

                    CUTTER
          I need to check the torpedoes.
          Where's Kitty?

INT. USS LEVIATHAN - CORRIDOR - DAY

Kitty carries Mister in her case carefully.

                    KEATING
          It's just down this corridor,
          over here. We're sharing the
          quarters. Kitty, I've heard a lot
          about you...I just wanted you to
          know that--it's an honor to bunk
          with you.

                    KITTY
          Thank you Emma, I'm afraid it's
          been awhile since I've been on a
          sub. I don't usually feel so
          well.

                    KEATING
          Well you are with the good
          doctor. I got plenty of drugs.

Keating opens the door.

INT. USS LEVIATHAN - KITTY'S ROOM - CONTINUOUS

They enter the room.

>   KEATING
>   It's small, but we've got our own
>   washroom.

She walks over to a very narrow door and opens it revealing a
small bathroom.

>   KITTY
>   Okay. Well...real small.

Keating sees Mister in the carrying case.

>   KEATING
>   And who have we got here? Does
>   the Captain know about this? I
>   was wondering about that case. It
>   kept moving around.

>   KITTY
>   I gave him the good news on the
>   way here. I couldn't leave her.
>   She's still a puppy. Her name is
>   Mister. Don't worry about the
>   Captain. I got her when our
>   Captain, Cutter, my former fiancé
>   decided not to show up to our
>   engagement party.

>   KEATING
>   He what?

Kitty takes Mister out. She BARKS and runs away.

>   KITTY
>   Quick, get her.

Mister bolts out the door. They both run after her.

INT. USS LEVIATHAN

The run down the corridors and through very large doors.

>   KEATING
>   I'll get her, I used to be a
>   track star.

>   KITTY
>   She's fast.

>   KEATING
>   Look, there she is.

Mister gets stuck at an ice hill where she keeps moving her
legs but getting nowhere.

                        KITTY
              Having problems honey.

Mister picks her up gently.

                        KITTY (cont'd)
              Let me get your apartment.

INT. USS LEVIATHAN - WEAPONS ROOM - DAY

Cutter and O'Brien walk into a room filled with nineteen
torpedoes. There are twelve Nitrogen, six attack and one
nuclear torpedo. All labeled dangerous.

The one nuclear torpedo is under a special secure lock and
key. Labeled, live nuclear weapon, use care.

The torpedoes are shaped like mini versions of the USS
Leviathan.

                        O'BRIEN
              These are my babies. Here's the
              liquid Nitrogen. Just
              percolating. Ready to go.

Large tanks containing liquid Nitrogen with steam forming
around them sit along side the wall.

We see signs that say, "HIGH VOLTAGE LASER LIGHTS IN USE,
WEAR GOGGLES."

                        CUTTER
              Where are the escape pods?

                        O'BRIEN
              Over here.

INT. USS LEVIATHAN - EMERGENCY ESCAPE ROOM - DAY

                        O'BRIEN
              These are the escape pods. One for
              one. One for two, and one for ten.

                        CUTTER
              Thirteen. Lucky number.

Cutter exits, turns and bumps into Kitty.

                        KITTY
              There you are. Come on, let's get
              something to eat. I'm starved.

Cutter starts to head in the wrong direction, Kitty corrects
him.

                        CUTTER
              Just kidding. I know which way it
              is.

INT. USS LEVIATHAN - MESS HALL - LATER

Mess, the ultimate use of space. Benches and fold away tables
fill the room. The CLANKING of silverware. The smell of good
hot food.

CHEF, 40s, happy to be cooking, overweight, Navy style.

The first big meal out to sea. Plenty of food. Chef is
sharpening his knife.

                    CHEF
          Thank you all for coming. Hope
          you enjoy. Today's menu is
          marinated roasted chicken breast
          quarter in thyme cognac sauce
          with spring morel mushrooms, pan
          sauteed sugar snap peas and
          asparagus, risotto cakes with
          wild mushrooms, lemongrass,
          saffron and red wine braised
          onions. Sorry no truffles, they
          wouldn't okay it in the budget.
          Step right up.

                    KITTY
          Really? You eat like this all the
          time?

Kitty grabs a tray.

                    CHEF
          That's right.

Emma gets utensils.

                    KITTY
          Hi, Emma.

                    KEATING
          Kitty. The meals on these boats
          are very expansive be careful.

                    KITTY
          We never got meals like this on
          the base.

Chef starts serving the Chicken.

                    CHEF
          Aye, men eat well and ladies,
          too.

Danish enters slowly, confused. Looking around.

                    CHEF (cont'd)
          Can I help you?

Danish stares at Chef.

                    DANISH
          What happened? I was visiting...I
          have to get off this boat before
          we leave dock.

                    CHEF
          Sorry, you are a little late,
          laddie. We've been out for five
          and a half hours. What you been
          sleeping?

Danish checks his watch. TAPS it gently.

                    DANISH
          Oh...I'm feeling sick. I need
          Dramamine. Doctor?

Kitty has Mister with her.

                    KITTY
          I have someone I want to
          introduce to you all. Hello,
          everyone. This is Mister. She
          only eats a little.

Mister BARKS.

Danish jumps back.

                    KEATING
          Let me help you.

                    DANISH
          Sorry.

Brick, Manello and Fenster enter.

                    FENSTER
          Well, hello all.

Around Fenster's neck is a SMALL RED KEY, very distinctive.

                    DANISH
          Five hours?

Cutter enters.

                    CUTTER
          Fenster put the red key away.
          Danish?

                    DANISH
          Captain. I'm not supposed to be
          here.

                    CUTTER
          Well. Use the side exit.

                    DANISH
          But.

                    CUTTER
        Not now.

Danish sits after taking a cup of coffee.

                    DANISH
        Coffee.

                    FENSTER
        With this red key...

Fenster takes the key off and puts it down on the table.

Just then Mister sees the key, BARKS, JUMPS, and SNATCHES the key with her teeth. She quickly sprints out.

                    KITTY
        Mister, come back...

Fenster pulls out her gun. Takes aim.

                    FENSTER
        I'll get her.

                    CUTTER
        Put that gun away.

Kitty, Cutter, Fenster, Brick, and Manello all run after Mister.

Mister runs down the hallway and around a corner. The group splits up.

Brick and Manello go down one hallway towards the weapons room.

EXT. USS LEVIATHAN - WEAPONS ROOM - DAY

They knock on the door...There is a glass viewing area that goes from opaque to clear. There is an intercom.

                    O'BRIEN
        How may I help you laddie? You're
        looking like the kid who lost the
        keys to the house.

O'Brien's face is right against the door.

                    BRICK
        We are looking for a small dog
        with a red key.

                    O'BRIEN
        A dog in this sub?

                    BRICK
        It's got the key, you know, the
        secured launch key.

                    O'BRIEN
        You been drinkin'?

                    BRICK
          I'll be back.

                    O'BRIEN
          Always open.

Camera pans down to reveal that Mister is standing next to
O'Brien. (A beat)O'Brien opens the door. Mister runs out
still with the red key.

INT. USS LEVIATHAN - HALLWAY - CONTINUOUS

Mister runs into the hallway carrying the chewed up red key.
The key which looked like metal is actually just shiny
plastic. Mister brings the key to Cutter. Cutter takes it.

Cutter hands the chewed up key to Fenster.

                    CUTTER
          Here. Try to hold onto it this
          time.

Fenster looks at the key carefully. Trying to figure out
which way it goes.

                    FENSTER
          Do we have a spare?

Cutter shakes his head.

                    CUTTER
          Can we finish eating now?

INT. USS LEVIATHAN - HALLWAY - DAY

NEWS BREAK

We zoom in on a live news broadcast.

INT. NEWSROOM - MAUI - DAY

From inside the small Hawaiian newsroom. Decorated in bright
flowers and palm trees.

The Reporter talks while the video of the Whirlpool is shown
from various angles and distances.

                    REPORTER
          ...this strange ocean current has
          turned into an extremely large
          vortex, a type of giant
          whirlpool. Which now threatens
          the entire Hawaiian chain. As you
          can see it has also been shifting
          its' positions as well having
          almost come ashore at some
          points. It is reported that a
          special Navy sub has been
          dispatched to the area. This is a
          live news update. A massive
          cleanup from the earthquake
          continues.

> REPORTER (cont'd)
> Many Hawaiians believe the
> Maelstrom Whirlpool is part of an
> ancient legend and like tsunami's
> very deeply rooted in culture and
> survival, Action News Reporting
> Live update.

INT. RESCUE HELICOPTER/COAST GUARD VESSEL - DAY

The sound of the HELICOPTER. The fog is heavy. We INTERCUT between two locations.

> DALLAS
> Quick...we can't hold.

The camera pulls back to reveal the Whirlpool, the enormous horizontal waves with their RUMBLING SOUNDS fill the air.

We go inside the helicopter.

> JENNY
> Did you see Rosco?

> DALLAS
> We're checking, hold on.

Inside Coast Guard Hawaii vessel.

> ALAN
> Rescue what is your ETA?

> DALLAS
> We are looking at six or seven
> minutes.

> ALAN
> Roger.

Jenny sticks her head forward.

> JENNY
> I lost Rosco.

Rosco surfing around the edge of the Whirlpool. Waving to the helicopter.

INT. RESCUE HELICOPTER - DAY

Jenny screams from the helicopter.

> JENNY
> Rosco!

Rosco can't stop and goes in DEEPER.

The Cadets hold back Jenny.

> JENNY (cont'd)
> Wait...

INT. RESCUE HELICOPTER - DAY

The air around the Maelstrom Whirlpool spins upwards.

                    JENNY
          I've got to rescue him. He's my
          only student. Give me a chute.

                    CAP
          No.

                    JENNY
          I'll jump. Give me the chute.

                    CADET
          Wait--let me get a line on you.

Jenny grabs the chute, surfboard and jumps out the open door.

EXT. RESCUE HELICOPTER - DAY

Jenny's chute opens. She SCREAMS all the way down.

                    JENNY
          Geronimo.

                    CADET
          Be right there.

The Cadet JUMPS in after Jenny. It's a long, long drop. They
land on the outer edge of the Whirlpool.

INT. RESCUE HELICOPTER - DAY

                    CAP
          Have you gone crazy? You let the
          cadet jump in. Damn. In that
          mess.

EXT. MAELSTROM WHIRLPOOL - DAY

The immense vortex with its' oceanic forces pulls them in
effortlessly.

                    JENNY
          Where is he?

The Cadet looks for Rosco with binoculars from inside the
raft. Jenny is sitting on the surfboard alongside.

                    CADET
          I can't see him...sorry.

INT. BETTY'S CONGAS - DAY

INTERCUT between lower, middle, & upper decks.

On the boat, the Popalisky family at their best, everyone
running around.

Lucas is looking through the underwater camera, lower deck, with Sofia.

Wilson is PUMPING up a raft, middle deck.

Mary starts unpacking sandwiches, sodas and chips.

Tom is on the upper deck playing the Congas.

Martin, the Captain, is STEERING the boat and watching football on a small flat screen TV, upper deck.

> MARY
> Captain Martin? Where are the whales? You promised.

> MARTIN
> We should see them in just a few minutes. Lucas?

Lucas is accompanied by Sofia.

> LUCAS
> Looks cloudy.

> SOFIA
> What about this fuzzy thing?

> LUCAS
> Might be.

> SOFIA
> What about the Hydrophone?

Lucas adjusts the headset. He listens carefully. He starts adjusting knobs and flipping switches.

> LUCAS
> In the background. I can hear them. Here...listen, they're singing. I'll put it on the speaker.

We hear the long deep melodic songs of the whales.

Martin points to a group of humpbacks.

> MARTIN
> Thar she blows.

> WILSON
> A whale...look a whale.

The Humpback takes a breach out of the Ocean. Slapping the water hard upon landing sending white foam everywhere.

> MARY
> Tom, where are you? Sofia?

                    SOFIA
             (she screams from below)
          I'm right here.

The fog rolls

                    MARY
          I am sorry--but this Mom is on
          vacation. Please. Is it getting
          foggy or what?

                    TOM
          I was just on my way.

He has stopped to watch football with Martin.

                    MARY
          Wilson be careful.

Wilson holds up the GPS.

                    WILSON
          GPS...see ya, does the work of
          parents.

                    MARY
          Wait. Sofia.

                    SOFIA
          I'm fine!

Wilson goes over to the edge of the boatwater with
his snorkel, mask and fins and jumps on top of a raft
and starts to paddle.

                    MARY
          Tom...what is it that you are
          helping with up there?

Tom is playing the Conga on the right side of the boat. The
sound moves to the bottom of the boat then through the water
to the whales who turn their bodies and heads to listen.

                    TOM
          I'm watching the game, playing
          congas with the whales, and
          looking for the Yankees score.
          Why?

                    MARY
          Are the whales singing back?
          Would you help keep an eye on
          your son?

                    TOM
          Sure. And they do sing back.

                    MARY
          We talked about this before.
          Remember honey. Sofia!

                    TOM
          I'll stand right here on the
          side, this way I can watch the
          game, play the congas and keep my
          eyes on our dear children. Just
          call me "Parent of the Year."

                    MARY
          Highly unlikely.

                    SOFIA
          Yes, mom. What?

                    MARY
          Just checking. Thank you.

                    SOFIA
          Ugh.

INT. BETTY'S CONGAS - DAY

                    MARTIN
          Lucas!

Martin YELLS pretty loud.

                    LUCAS
          (looking up) Captain...Dad. I
          think we need to change the
          heading.(To Sofia) Come on, let's
          go swim with the whales.

                    MARTIN
          Thanks. Let me look.

Martin looks over the instruments.

                    LUCAS
          Come on.

                    SOFIA
          Right now? I wasn't thinking of
          actually going in the water. This
          isn't a swimming bikini

Lucas motions for Sofia to join him.

                    LUCAS
          Forget about it. Let's go.

He jumps in wearing cutoffs and then she takes off her shorts
and jumps in. She's wearing a gold shiny bikini.

EXT. BETTY'S CONGAS - OCEANIC SWIMMING SPOT - DAY

A large Humpback whale rises from the water. Smashing down
with enormous force making a large wave. Sofia is pushed into
Lucas by the Whale's motion.

                    SOFIA
          ...this is just so amazing.

                    LUCAS
          Look it's the mother and calve.
          We won the bet.

                    SOFIA
          You bet on whales? My dad would
          like you.

                    LUCAS
          Well. Sort of.

The whale swims by, turning it's head out of the water and
taking a good look at the young couple. Then the calve swims
by doing exactly the same.

                    SOFIA
          They are so close. We touch them.

                    LUCAS
          Go ahead. They'll probably swim
          away. This current is weird.

                    SOFIA
          Whales...it's a family. Oh, it's
          a bunch of families.

Dozens of humpbacks swim by.

                    LUCAS
          Why do you think humpbacks are
          called humpbacks?

                    SOFIA
          Is this a sex quiz?

                    LUCAS
          Yes. No. 'Cause of the way they
          hump their backs when they swim.
          Like this.

Lucas demonstrates this up and down motion with his hand in
an arc.

                    LUCAS (cont'd)
          This is their annual route. They
          all pass by here. Every year. For
          millions of years.

While circling around Sofia and Lucas the whale SHOOTS out
some air and water from it's blow hole.

                    SOFIA
          Whales are so friendly. They know
          we are good people. They can
          tell.

                    LUCAS
          Bigger brains than humans. They
          were almost killed out of
          existence.

                    SOFIA
          People don't hunt whales.

                    LUCAS
          Just a few.

The camera PULLS BACK to reveal a procession of whales.

As we pull further back we see that Lucas, Sofia and the
whales, are all heading towards the outer rim of the
Whirlpool.

EXT. BETTY'S CONGAS - DAY

                    MARTIN
          Strange. But we won the bet.

Martin TAPS the gauges. The boat sways from side to side.

                    MARTIN (cont'd)
               (On the 2-way radio)
          Tam? Tammy? Tammy McFadden. Over.
          Tammy McFadden come in. Over.
          What the hell?

An AUTOMATIC SCROLL on one of the computerized weather
instruments prints out storm warning notices. The radio
interrupts with a warning. Bells ring.

                    ALAN
               (on the 2-way radio)
          This is Coast Guard--is this
          Betty's Congas--

                    MARTIN
          Yes. Coast Guard. This is Betty's
          Congas.

                    ALAN
               (on speaker)
          This is Coast Guard Rescue, you
          are heading right into a deep
          rapid depression. Immediately
          reverse to a new heading, over.

                    MARTIN
               (on the 2-way radio)
          What is it?

                    ALAN
               (on speaker)
          A whirlpool, a giant whirlpool, a
          MAELSTROM WHIRLPOOL. Like the
          world has never seen.

                    MARTIN
               (on the 2-way radio)
          A what? You're kidding? Who is
          this?

                        ALAN
          Coast Guard Hawaii. How many on
          board?

                        MARTIN
               (on the 2-way radio)
          Six on board. What happened to
          The Whalefinder?

                        ALAN
          The Whalefinder. Missing, we just
          don't know.

Alan on the radio, looking hard into the fog.

                        ALAN (cont'd)
               (on speaker)
          We'll start heading towards you--
          try and reverse direction. Over.

                        MARTIN
               (on the 2-way radio)
          ETA?

                        ALAN
               (on speaker)
          Twenty-five to thirty minutes.

                        MARTIN
          Great. Lucas? Where's Lucas?

EXT. PACIFIC OCEAN - BETTY'S CONGAS - SWIMMING SPOT - DAY

Lucas and Sofia have drifted further from the boat.

                        LUCAS
          Grab onto the raft.

                        SOFIA
          Okay, but we seem to be drifting
          further. Maybe we should head
          back.

                        LUCAS
          I know.

                        SOFIA
          I thought you did this all the
          time.

                        LUCAS
          Hey, I'm from Jersey, don't
          worry...we'll be okay.

INT. BETTY'S CONGAS - UPPER DECK - DAY

Martin looks out. First checks the radio, then the walkie
talkie, then grabs the mike. Then the cell. Lucas answers the
cell.

                    MARTIN
          Lucas. What the hell. Where are
          you?

Betty's Congas is caught on the edge of the Whirlpool.

                    MARTIN (cont'd)
          Lucas?

                    LUCAS
               (on cell)
          Dad? Where are you? I can't see
          anything. Hello?

The call gets disconnected.

Betty's Congas goes deeper into the crushing ocean vortex.

                    MARTIN
          I'm getting sucked inside this
          thing. Tom, Mary get ready to
          abandon ship. Take the raft, it
          holds four. Go quickly.

                    MARY
          Umm. Where are my children?

                    MARTIN
          On the water. Go. Quickly.

Tom and Mary grab the raft.

                    TOM
          Where?

The inflatable raft has a large gash on its side.

                    MARTIN
          It's no good. Wait I have an
          idea. Here...the Congas. Take
          them. Use them as lifeboats.
          Quick.

Martin runs over to the first Conga pulling the emergency
cord, releasing it into the Ocean.

                    TOM
          The Congas, great.

EXT. BETTY'S CONGAS - SWIMMING SPOT - DAY

                    LUCAS
               (on cell)
          Dad..Dad...DAD!

INT. BETTY'S CONGAS - UPPER DECK - DAY

We follow Martin as steers the boat towards Lucas and then
releases the second conga.

                    LUCAS
          Thanks Dad.

Betty's Congas gets pulled deeper into the edge of the Whirlpool. The boat swaying from side to side.

                    MARTIN
          I've never seen anything like
          this. Where did I put that last
          freakin' life preserver? Must get
          organized. Betty was right I'm
          going to die looking for a
          freaking life preserver.

Martin searches through drawers loaded with old gear.

INT. USS LEVIATHAN - CONTROL ROOM - LATER

The lights are flickering, there is steam in the air.
The room is hot. Everyone is slow from the heat.

                    CUTTER
          Hot. Hot as hell. Call room
          service.

Cutter TAPS on the wall thermometer.

                    EXEC
          There's something wrong with the
          AC. We have an incoming message
          from General Cherry. Hold on.

                    CUTTER
          Great.

On the screen, a simultaneous conversation with General
Cherry in Washington D.C.

                    CHERRY
               (on screen)
          Cutter...what is going on there?
          What's your ETA?

                    CUTTER
          Ninety minutes.

                    CHERRY
               (on screen)
          Well, What's taking so long?

                    CUTTER
          She's running hot.

                    CHERRY
               (on screen)
          Hot? You're underwater. Alright
          well...keep it going. Over and
          out.

Video transmission ends.

                    KITTY
          See. It's hotter in the center
          due to conservation of angular
          momentum. You know, it rotates
          slower at the outer edge.

They are staring at a monitor with the satellite image coming
forward. Then a 3-D projection.

                    KITTY (cont'd)
          It's hard to see with all this
          infernal fog. Look at the 3-D.
          It's a boiling cauldron in the
          middle.

                    BRICK
          How hot?

                    KITTY
          Maybe two hundred. Lots of cooked
          fish. Under pressure--it's a
          cooker.

                    CUTTER
          Better get this AC fixed. What's
          our position?

                    EXEC
          Here (points to the map) and
          three miles down.

                    CUTTER
          Under the Canadian ice pack.

All of a sudden a VERY LOUD SOUND is heard.

                    CUTTER (cont'd)
          Reactor three just kicked in.

Then ANOTHER SOUND is heard. A deep rumbling.

                    KITTY
          What's that sound?

                    SUBMARINE VOICE
          Reactor three has just failed,
          sorry.

                    EXEC
          We are in thick ice. See.

The Exec points to a monitor showing the various layers of
white and gray ice.

                    KITTY
          The ice used to be a lot thicker.
          A lot thicker. You know some day
          all this ice will be gone.

                    EXEC
          Let's just hope we can get
          through this now.

They look at a monitor showing a series of images of the
POV of the USS Leviathan boring through the ice.

                    EXEC (cont'd)
          We are still traveling at 250
          knots.

                    CUTTER
          Amazing. Slow it down.

INT. USS LEVIATHAN - CONTROL ROOM - DAY

They are listening to sounds from outside the sub.

                    CUTTER
          What's that sound?

                    SONAR
          Captain.

A loud BARK is heard.

                    CUTTER
          Recognize it?

                    SONAR
          Sounds like a dog bark. Nuclear,
          probably Russian. Maybe a Yorkie.
          That's Chinese music.

Inside the Russian sub they play CHINESE MUSIC. Which we can
hear.

                    SONAR (cont'd)
          No wait...something about this.

                    FENSTER
          It's Russian...Emperor's class.
          Been following us for hours.

                    CUTTER
          Is there any reason why you
          didn't say anything? Usually we
          like to share information
          especially on the same
          mission...it's kind of
          traditional. Evasive actions.

The USS Leviathan goes hard to one side.

                    FENSTER
          They are following our tunnel in
          the ice.

                    SONAR
          Incoming.

                    CUTTER
          Defensive actions.

The color of the lights change. The control board transforms into deep red lights and black dials.

> CUTTER (cont'd)
> Fire the torpedo, close the
> tunnel.

A torpedo FIRES and closes the tunnel for a moment and then the Russian sub reemerges.

> EXEC
> Captain?

> CUTTER
> Full ahead. No barking.

Mister BARKS.

> BRICK
> Send them a nuke.

> FENSTER
> Attaboy.

> CUTTER
> That's not a good idea.

> KITTY
> They won't catch us. Not in this
> mess.

The USS Leviathan continues through the tunnel, leaving crushed ice in its' path.

EXT. COAST OF MAUI - WILSON'S RAFT - LATER

Wilson has been drifting.

> WILSON
> (talking out loud)
> Great. The Captain at sea. Who
> needs lousy parents anyway. GPS.
> I'm set. Huh, wait, what the
> heck? What's going on? I broke
> it.

Wilson holds up the GPS unit...but something is wrong with it, he shakes it, presses some more buttons...it goes on and off. It shows a brief satellite image of the Whirlpool.

The camera PULLS BACK to reveal that Wilson is heading right into the vortex.

> WILSON (cont'd)
> (on phone)
> Hello, anyone there? Hello? Mom?
> Dad?

Wilson keeps switching frequencies.

> ALAN
> This is Coast Guard Hawaii.

                    WILSON
          Coast guard. Hold on...(the GPS
          flashes on and off) Latitude 18
          degrees 31 minutes, 53 seconds
          north, longitude 154 degrees, 36
          minutes, 10 seconds west. I'm on
          a raft. My name is Wilson.

Wilson looks around.

                    ALAN
          How old are you?

                    WILSON
          Twelve.

                    ALAN
          Young man...hold on.

Wilson holds onto the raft. The whirl is pulling him in.

                    WILSON
          I think the current is getting
          stronger.

                    ALAN
          There's a big rescue helicopter
          on the way. Stay calm.

The camera FOLLOWS Wilson around the Whirlpool.

INT. USS LEVIATHAN - CONTROL ROOM - LATER

                    CUTTER
          Let's see. Where are we?

                    FENSTER
          Are you actually lost?

A holographic series shows their paths.

                    KITTY
          He's always getting lost. Even
          under water. Are you actually
          asking for directions? This is
          our original path. This is the
          new course.

Now the holograph shows the Maelstrom Whirlpool.

                    CUTTER
          We are supposed to cut through
          the sediment.

                    KITTY
          Right, we have to figure out
          which way to approach.

Fish hands Cutter a map of the area, he flips it around a
couple of times.

                    FISH
          Here Captain. This is the area
          around the Whirlpool. The
          question is... where is all the
          water going?

                    FENSTER
          Who cares?

                    KITTY
          Maybe a giant subterranean sea?

                    FISH
          Really?

                    EXEC
          Captain, we have an incoming
          message from the White House.

                    CUTTER
          Put it on. The President.

INT. NATIONAL SECURITY COUNCIL - SITUATION ROOM - DAY

Via satellite. CHIEF OF STAFF, 50s, Defense Department. In
the middle of a conversation with the General.

                    CUTTER
          Hello, General.

                    CHERRY
          At the current rate, they should
          be there in about six hours.
          Cutter how's it shaking?

                    CHIEF
          What about that Russian sub?

                    CUTTER
          I think we took care of that. No
          barking.

There is LAUGHTER in the room. Mister BARKS.

                    CHIEF
          I need to inform The President of
          your recommendation.

INT. THE WHITE HOUSE - OVAL OFFICE - DAY

The office of PRESIDENT, 50s. There is a hair dresser and
manicurist working.

                    PRESIDENT
          Well, Cherry, this is your bag,
          tell me what you think. (softly,
          to hairdresser) Not too much off
          the top, and please use clear
          polish this time. This could cost
          me votes if it goes bad, General.
          Cutter, what do you think?

                    CUTTER
          I'm sure we can take care of
          this, Mr. President.

Mister BARKS.

                    PRESIDENT
          You've got a dog on that Sub?
          Great idea but where do you walk
          him?

                    KITTY
          Shush. No, Mr. President.

                    PRESIDENT
          Well, as long as I can count on
          both your votes. Those news guys
          they keep trying to tie this
          migraine whirlpool to global
          warming and my administration.

                    KITTY
          It's the Maelstrom Whirlpool, Mr.
          President.

The transmission is BREAKING UP.

                    PRESIDENT
          What? Great, Well, glad we can
          agree on this.

                    CUTTER
          But...

The President CLICKS OFF the transmission.

                    CHERRY
          That's the plan. Over and out.

                    CUTTER
          Well, there you have it. I'm
          going to return to my quarters
          would you like to join me?

                    KITTY
          What do you think Mister? The
          Captain's quarters. Don't play
          dead, that's not funny.

Mister plays dead.

INT. USS LEVIATHAN - HALLWAY - LATER

The Captain's Quarters is the best quarters on the sub.
Complete with flat screens, leather couch and a small bar.

Kitty and Cutter enter, Mister jumps into the chair.

                    CUTTER
          Please enter. Music? Let me see
          if I remember.

The bachelor in his pad.

Kitty sits. Cutter opens a HIDDEN COMPARTMENT in one of the walls.

>   CUTTER (cont'd)
>   Cocktail perhaps?

>   KITTY
>   We are in the middle of a very
>   important mission. What have you
>   got?

He takes out a bottle of CAPTAIN COOK'S RUM.

>   CUTTER
>   Captain Cook's Rum?

>   KITTY
>   Are you kidding? Where did you get
>   that?

>   CUTTER
>   It was a going away present.

>   KITTY
>   From a girlfriend?

Kitty gets up and Cutter sits down.

>   CUTTER
>   You can't be jealous.

>   KITTY
>   And why not?

>   CUTTER
>   Because.

Kitty walks over.

They are eye to eye.

>   KITTY
>   What about some commitment?

Kitty turns away.

>   CUTTER
>   You know I'm committed. I'm sorry.

>   KITTY
>   Really?

Cutter moves in closer.

>   FISH
>   (on intercom)
>   Captain, sorry to disturb you.

They both freeze.

                    CUTTER
          No problem.

                    EXEC
                 (on intercom)
          We're picking up that Russian sub
          again.

                    CUTTER
          I'll be right there. You want to
          wait here?

Cutter starts to exit, Kitty follows immediately after.

                    KITTY
          Like last time?

                    CUTTER
          It was an honest mistake.

                    KITTY
          You went on a three month tour.

INT. USS LEVIATHAN - HALLWAY - LATER

They walk through the narrow hallway, alternating leads.
Mister following.

                    CUTTER
          I was assigned.

                    KITTY
          You volunteered.

                    CUTTER
          It was a new sub.

                    KITTY
          I needed you.

They continue walking through the ship, arguing.

                    CUTTER
          I told you...I'd be back.

                    KITTY
          You missed our Engagement party.

                    CUTTER
          ...when you're out in the field
          you don't rush back to the
          office. I had to go to sea. I got
          to go.

INT. USS LEVIATHAN - CONTROL ROOM - DAY

Brick, Kitty, Cutter, Fish and the Sonar operator are in the
Control Room.

                    BRICK
          I would never have missed our
          Engagement party.

                    BRICK (cont'd)
Most likely it would have been an
early evening especially the
wedding night. That would have
been real good.

                    KITTY
Really. Is that so? I bet it
would have been quite a
Honeymoon.

                    BRICK
You got that right. Miss.

                    CUTTER
Don't encourage him.

                    BRICK
Please...I'm sensitive.

                    KITTY
You're cute but just twenty years
too young. I like old men.

                    CUTTER
Thanks a lot.

                    SONAR
I've got a fix on that sub again.

They all listen to the SONAR SYSTEM, announcing the position.

                    SONAR (cont'd)
It looks like we have an incoming
message, and if I'm not mistaken
this is going to be in Russian.

                    RUSSIAN VOICE
                (on speaker)
Unidentified wessel, identify
yourself. This is Russian
territory. Did you not see our
special Russian underwater red
neon flag?

A screen shows a Russian flag underwater.

                    SONAR
Captain?

                    CUTTER
How long till the edge of the
Whirlpool?

                    SONAR
Current heading puts us at the
epicenter in twenty five minutes.

Cutter referencing the maps on the table.

                    CUTTER
Lose this sub...let's go
underground.

                    CUTTER (cont'd)
          We're going to approach north of
          Maui then go 45 degrees down
          through the sediments to the
          crust layer and then into the
          magma. Everyone ready? Let's
          project this course.

A series of Holographic images outline the
course.

                    CUTTER (cont'd)
          Bow 20 degrees, stern 15 degrees.
          Let's give them something to
          think about. Increase our speed
          to two hundred knots.

                    SONAR
          Incoming torpedo.

Warning lights and bells go on and off.

                    CUTTER
          Hold on everyone. Evasive action.

                    BRICK
          Hell with that. Send them a
          tomahawk.

                    CUTTER
          Fire six. Ready the Nitrogen.

                    MANELLO
          Hot dog.

                    CUTTER
          Prepare tube five.

INT. USS LEVIATHAN - WEAPONS ROOM - LATER

O'Brien and Fish are in the process of filling the tube with
liquid Nitrogen. We see monitors indicating a variety of
levels.

                    FENSTER
          Careful.

                    O'BRIEN
          This, this is nothing. I once had
          to load pure nitroglycerine
          during a hurricane with a bad
          case of the hiccups. In a
          submarine too. (hiccups)

On his last word O'Brien immediately begins to HICCUP.

O'Brien and Fenster are carefully pouring the liquid Nitrogen
when all of a sudden it spills all over the floor, instantly
it turns white, freezing everything it comes in contact with.

                    O'BRIEN (cont'd)
          Look out. Spill.

                    FENSTER
          Don't let it hit the electrical
          panel. Here grab this wood.

Fenster throws a block of wood at it. Which freezes, fractures
and then EXPLODES.

                    O'BRIEN
          Glad, we got that out of the way.

INT. USS LEVIATHAN - CONTROL ROOM - CONTINUOUS

Cutter screams into the microphone.

                    CUTTER
          Can you please tell me what the
          hell is going on down there? Fire
          the attack torpedoes.

An elaborate series of steps occur within the lighting of the
weapons room.

The lasers heat up producing a series of light beams. Steaming
the air it passes through.

The torpedo gets LAUNCHED. We see it go through layers of
silt and then it explodes near the Russian sub.

                    SONAR
          Captain, a near hit, Sir.

Everyone in the control room CHEERS.

                    CUTTER
          That's one quick ass torpedo.
          Let's work on getting into
          alignment.

                    EXEC
          Five minutes to the zone.

                    CUTTER
          What's it look like?

On the monitors.

                    EXEC
          Deep and dark. Here. It's got
          rotating currents reaching all
          the way to the bottom of the
          ocean.

Different images of the Whirlpool from several angles.

                    KITTY
          The Whirlpool is rotating at
          about 200 knots. We don't want to
          get caught in that. But then
          sometimes it stops completely and
          then starts up.

                    CUTTER
          Dive through the crust.

The Leviathan's red interior lights turn ultraviolet.

                    CUTTER (cont'd)
          Now, let's see.

The gauges indicate overheating in progress.

                    CUTTER (cont'd)
          Ventilation?

                    EXEC
          On maximum.

                    MANELLO
          Hotter than Brooklyn in August.

                    CUTTER
          I'll run sea water thru the auxiliary
          vents to the cooling systems.

The Exec stops him and then starts to flip switches and turn
dials checking different temperature readings.

                    EXEC
          Venting is different in this sub.
          But I think we can run it from
          the stern under the engine room.

                    KITTY
          Find the edge of the Maelstrom.

                    SONAR
          It's about ten miles--heading 270
          degrees.

INT. USS LEVIATHAN - WEAPONS ROOM - DAY

O'Brien BANGS his hand hard against the console, then a green
light goes on.

                    O'BRIEN
          Number two is ready.

                    CUTTER
          Fire torpedo.

INT. USS LEVIATHAN - CONTROL ROOM - DAY

                    SONAR
          We missed.

                    CUTTER
          Turn her around.

                    KITTY
          It's no good we have to approach
          from the other side, against the
          current.

                    CUTTER
          Increase speed.

                    EXEC
          Captain, we're being pushed.

The USS Leviathan rises from the magma and goes towards the
Whirlpool, struggling in the current.

                    CUTTER
          Damage report.

                    EXEC
          Radiation above normal but just
          below critical.

                    CUTTER
          How many horsepower does this
          DSRV push? Hey, stop that Mister
          not in the sub.

Mister urinates.

                    EXEC
          Very different sub. About seventy five
          million. I'll check the engine room.

                    CUTTER
          You said the Whirlpool produces
          seventy million horsepower.

                    KITTY
          That's not exact. What are
          suggesting? Wait! You're not
          suggesting that we try and outrun
          it. Are you? I mean--it's like
          trying to outrun a tornado from
          the inside!

                    CUTTER
          All ahead full.

We see the USS Leviathan try to level out, but it's losing
ground.

INT. USS LEVIATHAN - ENGINE ROOM - DAY

The engine room. Where radioactivity from the nuclear
reactors is used to make steam that run the turbines that
power the sub.

The walls are filled with electronics and video screens
showing temperature sensors, pressure measurements and
radioactivity levels. All reading critical levels.

                    EXEC
          Captain, we can't keep this up.
          We are still being swept backward
          by the current. Systems critical.

The USS Leviathan fights the currents around the Maelstrom
Whirlpool

EXT. USS LEVIATHAN - DEEP OCEAN - DAY

POV from the sub.

We see smashed up boats passing in front of them.

INT. USS LEVIATHAN - CONTROL ROOM - DAY

                    CUTTER
          Hope they got away.

Bodies float by. Then fish and other marine life. Some are
swimming against the current, others are swimming upside down
and with the current.

                    KITTY
          Oh, my.

                    BRICK
          Shit.

                    MANELLO
          Lord.

                    EXEC
          Steady as she goes.

                    CUTTER
          Again.

                    KITTY
          Let's back out of here.

There is a moment of SILENCE.

                    EXEC
          At maximum...

                    CUTTER
          Let's get to the bottom of this
          tunnel of hell.

                    EXEC
          Forty five degrees down. Hold on.

They all grab something.

                    HELMSMAN
          Forty five degrees down rudder.

Emergency lights flash on and off.

                    EXEC
          CUTTER. We're going up!

The USS Leviathan veers to the left and then starts to go up.

                    CUTTER
          Up? How can we go up?

                    EXEC
          We're going up.

                    CUTTER
          Everybody hold on.

The USS Leviathan is moving straight up. The result of the
forward push of the engines and the counter-clockwise
rotation of the Whirlpool.

                    EXEC
          Shit!

                    SONAR
          Depth, twenty two hundred feet,
          two thousand feet, fourteen
          hundred feet, one thousand feet.

                    CUTTER
          Prepare to surface, maintain
          speed. Try to veer out of the
          Whirlpool.

                    KITTY
          The current is still too much.

                    EXEC
          She won't go.

                    SONAR
          Depth now five hundred feet, now
          two hundred feet.

                    CUTTER
          Ready to be airborne. Here we go!

All of a sudden, the USS Leviathan leaps out of the Ocean
like a whale taking a much needed breath.

EXT. RESCUE HELICOPTER - DAY

Inside the front cabin of the helicopter they quickly turn to
see...

                    CAP
          Dallas, what the hell was that?

                    DALLAS
          That's either one strange boat or
          the biggest and ugliest damn fish
          I've ever seen.

EXT. RESCUE HELICOPTER - DAY

Inside the rear cabin of the rescue helicopter. They lean
out.

                    JENNY
          I wanna get me one of those.

EXT. COAST OF MAUI - WILSON'S RAFT - DAY

Wilson inside the raft, holding on tight.

                    WILSON
          Coast Guard? Hello?(into the cell
          phone)

                    CAP
          Hello, this is Air Sea rescue.
          Son, we'll be right there. I can
          see you. Get the basket ready.

The lights and electricity start to go on/off, small bursts
of SPARKS come from behind the panels of the helicopter.

                    DALLAS
          Okay. Lower the basket. Grab the
          boy.

The metal basket is swaying back and forth.

                    DALLAS (cont'd)
          Hold on.

EXT. RESCUE HELICOPTER - LATER

We cut to Wilson being brought on the Helicopter.

                    WILSON
          We have to find my parents.

                    CAP
          Right. Cadet, you reading this?

                    CADET 2
          Reading you Cap.

INT. RESCUE HELICOPTER - DAY

Inside the helicopter.

                    CAP
          Is that a drum floating over
          there?

Cap spots the Congas from Betty's Congas.

                    WILSON
          A conga. My parents, over there.
          In the Conga.

Close shot of the two Congas floating in the ocean-one with
Tom and Mary and the other with Lucas and Sofia.

                    CAP
          Let's lower the bucket although
          I'd say drop a guitar down there
          and they'd have a band.

                    CADET 2
          Lowering the basket. Please stay
          inside the drum until the basket
          comes to you.

                    CAP
          It's a Conga.

                    MARY
          Tom, grab the rope. Look how big
          that nasty whirley thing is. My
          lord.

Tom and Mary get pulled up. Mary turns and sees the enormous
Whirlpool, with thick fog circling above it

                    DALLAS
          That's ok, just hold on. You guys
          are next. It's close to two
          hundred miles in diameter now.

                    LUCAS
          Grab the rope.

Lucas and Sofia with their arms wrapped around each other get
towed up in the basket.

                    LUCAS (cont'd)
          Kind of cozy like this.(beat)We
          have to find my father.

Martin has been steadily drifting towards the edge of the
Maelstrom Whirlpool.

                    DALLAS
          Martin?

                    LUCAS
          My Dad. He still might have the
          satellite phone.

INT. BETTY'S CONGAS - UPPER DECK - DAY

Betty's Congas is caught in the Maelstrom Whirlpool.

                    MARTIN
               (on cell)
          Coast Guard. This is Captain
          Martin of Betty's Congas. I am
          trying to pull out of this
          current. I seem to be circling.

We see Betty's Congas caught in the enormous force of the
Maelstrom Whirlpool.

INT. COAST GUARD VESSEL - DAY

                    ALAN
          This is Coast Guard Rescue. What
          is your position?

EXT. BETTY'S CONGAS - DAY

                    MARTIN
          Hold let me see if I can get a
          reading. Oh, no look at that wave
          coming my way. Hurry up, Coast
          Guard.

Just then a large wave knocks Martin from the boat and right
into the water.

                    MARTIN (cont'd)
          Great, not even going down with
          my boat. Hey, hey, what the heck
          is going on? Flippers who can
          this be?

Martin is drowning when all of a sudden a whale pushes him
up, and towards the rescue spot.

                    MARTIN (cont'd)
                 (to the whale)
          I owe you one.

The whale SNORTS loudly. Martin and the whale have an eye to
eye moment. Martin grabs the basket. They lift him up to the
helicopter.

INT. RESCUE HELICOPTER - DAY

                    CADET 2
          Got a friend down there.

                    MARTIN
          They sure love the Congas. So
          long Betty.

Martin salutes.

                    DALLAS
          Of course. Hold on.

INT. USS LEVIATHAN - CONTROL ROOM - DAY

The USS Leviathan jettisons out of the ocean. The crew
becoming weightless for a moment. They hold on tightly.

                    CUTTER
          Brace for surface break. We are
          airborne.

                    EXEC
          Captain, we're really airborne.

                    SONAR
          That's a first for me.

                    EXEC
          Our speed is one hundred twenty
          five knots.

                    CUTTER
          An airborne DSRV, the Air Force
          is gonna want one of these.

                    EXEC
          Captain, we're about three
          hundred feet above sea level.
          Brace for impact. This baby is
          going to hit hard.

                    KITTY
          John.

Kitty and Cutter are together, arm in arm.

EXT. USS LEVIATHAN - DAY

The USS Leviathan PLUNGES back into the ocean, sliding
quickly through the water.

INT. USS LEVIATHAN - CONTROL ROOM - DAY

                    EXEC
          Captain...we're going down very
          fast.

                    CUTTER
          Reverse direction.

                    EXEC
          We're going in.

                    SONAR
          Reading depth: 600 feet, 900
          feet, 1400 feet, 2000 feet, 2400
          feet.

                    CUTTER
          Prepare to go underground.

The USS Leviathan carves its way into the earth, breaking
through the seabed.

                    EXEC
          Reading minus two hundred, six
          hundred, fourteen hundred, twenty
          two hundred, twenty six hundred.

There are SOUNDS coming from the hull area as the stress from
the outside pressure increases.

                    KITTY
          What is that sound?

                    CUTTER
          Equalizing.

                    EXEC
          Captain, stabilized.

                    FENSTER
          There's no time for the Nitrogen.
          We've got to set off the other
          solution. Captain, now.

                    CUTTER
          Listen I'm NOT nuking the ocean.
          Now calm down. You got that?

                    EXEC
          Captain, twenty three miles from
          center.

                    FISH
          Here, the latest satellite data.

We see images of the Maelstrom Whirlpool, shifting around
from side to side.

                    KITTY
          Two hundred miles across and a
          mile and a half deep. Where's
          Cutter?

INT. USS LEVIATHAN - WEAPONS ROOM

Brick and O'Brien are fighting. Brick has a heavy wrench,
O'Brien, a broom stick. Cutter throws open the door.

                    O'BRIEN
          I'll kill you.

                    BRICK
          You asked for it..

Cutter breaks them apart.

                    CUTTER
          Have you guys gone crazy? You
          can't fight in here, it's the
          Weapons room.

Cutter holds back O'Brien.

                    CUTTER (cont'd)
          Stop fighting. Shake. That's an
          order.

O'Brien and Brick reluctantly shake their bodies.

                    CUTTER (cont'd)
          Not your bodies, your hands.

Brick exits. As he walks he takes out a crumbled sheet of
paper torn from a book with "Top Secret" Nuclear Emergency
Detonation procedures written on it. He walks into a store
room.

The sub's ELECTRICAL SYSTEMS continue to go on and off.

We see a panel indicating the Nitrogen filling up in each
tube.

                    EXEC
                (on speaker)
        Captain, four miles away.

                    KITTY
                (on speaker)
        How's it going down there?

                    MANELLO
        Ready.

                    CUTTER
        Fire the lasers.

We see a beam burn through a section of the magma, then
again, then again, till the frequency is so high that a
tunnel is created.

                    CUTTER (cont'd)
        Fire tubes one, two and three.
        I'll be right there.

The torpedoes take off at an incredible speed.

EXT. USS LEVIATHAN - MAGMA - DAY

Number one torpedo explodes from a magma intrusion of hot
liquid rock.

INT. USS LEVIATHAN - CONTROL ROOM - DAY

                    SONAR
        Number one has detonated.

                    CUTTER
        Hold steady.

                    KITTY
        We should have gotten closer.

POV of the torpedoes boring through magma in a multicolored
three dimensional imaging system.

                    SONAR
        Number two is caught in the
        current. It's heading back
        towards us.

                    KITTY
        Fire the laser at it.

                    EXEC
        Captain?

                    CUTTER
        Just fire the laser.

                    EXEC
        Fire the laser.

                         O'BRIEN
                       (on speaker)
                    Firing the laser.

EXT. USS LEVIATHAN - MAGMA - DAY

We see the laser bore a tunnel which goes right by the
torpedoes. The torpedoes turn and follow the tunnel.

INT. USS LEVIATHAN - CONTROL ROOM - DAY

                          KITTY
                    Do you always do that?

                         CUTTER
                    Somehow I think it's only when
                    you're on-board.

                          KITTY
                    I thought I was good luck. You
                    tell him Mister.

Mister BARKS.

                          SONAR
                    It's eight hundred feet, six
                    hundred feet, four hundred feet.

                         CUTTER
                    Full right rudder maximum speed.

We see the torpedo pass right by without detonating.

                          EXEC
                    Number three is on target.

                          SONAR
                    ...one hundred feet, fifty feet,
                    twenty feet. It's dissipating.

We see an image of the torpedo on sonar turn white, then
start to dissipate.

                         CUTTER
                    Look. It's not enough Nitrogen,
                    we need more. How many gallons in
                    the bow?

                          EXEC
                    We're running one hundred
                    thousand gallons as the coolant
                    for the three super reactors. The
                    Navy is not going to be happy
                    about you scuttling another ship.

                         CUTTER
                    We just have to.

                          KITTY
                    Must save the planet.

                    EXEC
          Ship expendable.

                    CUTTER
          Exactly.

Kitty is looking for Mister.

INT. USS LEVIATHAN - NITROGEN TANKS - DAY

The Nitrogen tanks are surrounded by a dense fog. Fresh white
ice has formed all around them. Cutter tries to push the
tanks closer together.

                    CUTTER
          We've got enough Nitrogen to
          freeze the ship and stop this
          Maelstrom Whirlpool.

                    EXEC
          Someone has to open all the
          valves and take the one man pod.

Kitty and Cutter push the tanks together while looking for
Mister.

                    FENSTER
          I'll do it. Here.

Fenster helps them move the tanks.

                    MANELLO
          No, I'll do it.

                    EXEC
          Sir, I volunteer.

                    CUTTER
          Denied, it's my call. Move
          quickly. Evacuate everyone.
          Abandon ship. Load the escape
          pods. That's an order.

                    KITTY
          You can't do it. There's a better
          way.

                    CUTTER
          Let's get to the tanks. Sound
          alarms.

                    EXEC
          I'll see you topside.

                    FISH
          I'm going with him, that is
          unless, you want me to stay.

                    CUTTER
          No. Go. That's an order.

The boat ALARM starts to sound loudly.

INT. USS LEVIATHAN - TEN PERSON ESCAPE POD - LATER

The large, ten person escape pod, lights on, ready to go.
With Danish, Chef, Keating, Fish, Fenster, Manello, Royal,
Planesman, Helmsman, and Fanucci.

                    DANISH
          What's happening?

                    EXEC
          Hold on tight. Launching pod.

The ten person escape pod is jettisoned into the ocean.

INT. USS LEVIATHAN - HALLWAY - LATER

                    KITTY
          Mister?

Kitty starts to look around for her dog.

They are running through the USS Leviathan.

INT. USS LEVIATHAN - COOLING SYSTEM ROOM - DAY

The room is a maze of TUBES AND PIPES. The main wall contains
all the controls for the entire sub.

                    KITTY
          This valve?

                    CUTTER
          No. That's the vertical outlet.

                    KITTY
          Where's Mister?

Kitty walks down another hallway. Calling out Mister's name.

                    KITTY (cont'd)
          Mister? Mister? Where are you
          honey?

Cutter is adjusting the valve.

                    CUTTER
          This one.

Liquid Nitrogen begins to flow from the pipes freezing
everything it comes in contact with.

                    CUTTER (cont'd)
          Time to leave. Kitty, Where are
          you? Mister? Damn Dog. Under the
          pipe. Now no nipping at me. Ok?
          Please. That's it...be nice.

Mister is hiding under a pipe. And for the first time Cutter
picks him up.

                    CUTTER (cont'd)
          Mister, I found you. Let's go.
          Now we got to find Kitty. Kitty!

Mister barks.

INT. USS LEVIATHAN - HALLWAY - DAY

                    CUTTER
          Here Kitty, here kitty. Where are
          you?

Cutter runs through the boat with Mister--as he does we can
see whole sections freeze as Nitrogen flows over everything.

INT. USS LEVIATHAN - ESCAPE PODS - DAY

Cutter runs for the two person escape pod but it is gone. All
that's left is the single person pod. Small, cigar shaped.

                    CUTTER
          WHAT? Who could have the two
          person pod?

                                                    CUT TO:

INT. OCEAN - TWO PERSON ESCAPE POD - CONTINUOUS

                    BRICK
          I do.

Kitty is laying face up in the passenger side of the pod.
Only one set of controls. She reaches for the controls.

                    KITTY
          Cutter! Help! He has me in here
          with him and the nuclear bomb.

Cutter opens the hatch for the single pod.

                    CUTTER
          Brick, listen son, stop that
          right now. You are to turn back
          immediately...

INT. TWO PERSON ESCAPE POD - CONTINUOUS

Brick has the nuclear bomb all tied up around him.

                    BRICK
          No way. I'm going to be the hero
          and Kitty is just my safety to
          make sure you don't try and stop
          me.

                    CUTTER
          Why would I do that?

                    BRICK
          'Cause I got your girl.

Cutter gets in the one person escape pod. Almost forgets
Mister but then grabs him. Puts on his helmet, shuts the
hatch and jettisons himself into the ocean.

                    CUTTER
          She's not my girl.

                    KITTY
          Cutter, he's not fooling around.

                    CUTTER
          Brick you're nuts. Turn back
          immediately. We've got a better
          plan.

We see on radar, the other pod moving quickly.

                    KITTY
          Brick, come on now. Listen don't
          do this. Let's head back.

Brick has a gun.

                    BRICK
          Shut up. This is my mission. My
          chance to save the planet

                    CUTTER
          Come on Brick, you don't want to
          hurt Kitty. Do you?

Mister barks.

The two person pod races towards the Whirlpool, followed by
Cutter in the other pod.

Kitty is desperately looking for something to distract Brick
with when she notices something in her pocket. The red key
toy.

                    BRICK
          What's that sound.

She squeezes it hard, producing a very loud squeaky noise
distracting Brick, she grabs his gun.

                    KITTY
          Cutter, I've got the gun.

                    CUTTER
          What? I'll be right there.

                    BRICK
          Don't think you can stop me.

                    KITTY
          No. Damn.

Brick hits the button, the red digital clock reads, three
minutes.

                        CUTTER
          I'm almost there, listen Kitty,
          There is a way to do this. I've
          done it before. Get your mask on.
          Hit the emergency release switch,
          I'll be right next to you. Upside
          down, blowing out air. You move
          right in. I'll be there in two
          minutes. We trick the airlock.

                        KITTY
          Can you go a little quicker? Brick
          please turn off the bomb.

                        BRICK
          Go. You go. I'll be just fine.
          Thank you and have a nice day.

Brick looks terrible. All wrapped up with wire. Sweating.

                        KITTY
          It would have been a great
          engagement party. Here.

She gives him a kiss on his cheek.

                        CUTTER
          Now!

Kitty hits the emergency switch, blasting out her half of the
escape pod. The other remains dry. She blasts herself right
into the other pod.

Cutter grabs her and scoops her up into his arms.

INT. USS LEVIATHAN - SINGLE PERSON ESCAPE POD - DAY

A tight fit. They are wrapped around each other like a
pretzel.

                        KITTY
          I'm freezing. We've got to stop
          meeting like this. Mister. Thank
          you darling for rescuing me and
          Mister.

Mister barks from inside his case.

Kitty gives him a kiss.

                        CUTTER
          Wait. Look, there he is.

On the control board, we see a second BLEEP.

INT. TWO PERSON ESCAPE POD - DAY

We INTERCUT between the single person pod(Kitty and Cutter)
and the two person pod(Brick).

                        BRICK
          No way. My job. My job.

                        KITTY
              Turn the bomb off.

Kitty and Cutter re-adjust their positions in the pod.

                        BRICK
              Kitty, I can't.

                        CUTTER
              Listen...It won't help. Turn the
              bomb off. We sent the sub loaded
              with Nitrogen into the eye of the
              storm. You are already a hero.
              You'll get at least one or two
              metals son and heck probably a
              promotion.

                        BRICK
              I'm kind of tied up here.

Brick still has wires around himself. The clock is ticking.

The two man escape pod starts to heat up.

                        KITTY
              He's going into the bottom of the
              whirlpool.

Brick gasps for air. The temperature is rapidly rising.

                        BRICK
              It's getting hot. I can't breath.

                        CUTTER
              Steer up...You're going the wrong
              way, you'll get caught in the
              current.

                        BRICK
              I can't move...it's stuck. No.

The two person escape pod BURSTS, the bomb detonating.
Sending shock waves in all directions.

INT. NATIONAL SECURITY COUNCIL - BAD SITUATION ROOM - DAY

The room is filled with military and civilian personnel
watching very large screen monitors of the Maelstrom
Whirlpool.

                        CHERRY
              No one is answering.

                        FARNSWORTH
              Less than an hour before it hits
              the coast.

                        CHERRY
              We can still fire a torpedo from
              the Utah.

                    FARNSWORTH
          A torpedo would never get there.
          We are going to have to call fora
          mass relocation. What? What
          explosion? Wait. We are receiving
          a report of a small nuclear
          explosion.

We see a screen of the Whirlpool with an analysis of water
speeds. Reading 260 knots in the center.

                    CHERRY
          Look at that explosion. I thought
          he wasn't going to use it.

INT. SINGLE PERSON ESCAPE POD - DAY

                    CUTTER
          Here comes the shock wave. Hold
          on.

                    KITTY
          Oh, no. I hate subs. I told you I
          did. I should never have come
          with you. I knew it.

                    CUTTER
          I wonder what the USS Leviathan
          looks like.

INT. USS LEVIATHAN - CONTROL ROOM

The walls and floors of the control room freeze.

We see the USS Leviathan bore through the magma. Moving
towards the center of the Maelstrom Whirlpool.

INT. SINGLE PERSON ESCAPE POD - CONTINUOUS

                    CUTTER
          Something's wrong. We're caught
          in the thermohaline circulation
          current.

We see them get pulled down lower. The BELLS AND WHISTLES
start to go off. They struggle to move the navigational
controls.

                    KITTY
          What? John. Oxygen low. Where's
          my ring?

Computer ANNOUNCES, "Oxygen low."

                    CUTTER
          Don't say goodbye just yet. Umm.
          It's on the sub.

                    KITTY
          No. I loved that ring. I'm sorry.
          And Mister, I shouldn't have
          brought you baby.

Computer ANNOUNCES, "Battery low."

Cutter talks really fast.

                    CUTTER
          Battery low. Look if it's any
          consolation, I really did want
          to marry you. I'm sorry.

                    KITTY
          You are a real Captain? Right?
          Can't you marry us now?

                    CUTTER
          I never thought of it. I guess
          so. I mean we're never going to
          get out of this.

Cutter smiles, not believing this at all.

                    KITTY
          I can hardly breath. Carbon
          dioxide levels are high.

Computer ANNOUNCES, "Carbon dioxide levels high."

                    CUTTER
          Do you Kitty Honeywell take me,
          Captain John Cutter, to be your
          husband? What a second. We can't
          go one with this... it's not
          valid.

                    KITTY
          Why not?

                    CUTTER
          We need a witness.

                    KITTY
          How about Mister?

                    CUTTER
          A dog? I'll have to check into
          this, but alright.

                    KITTY
          I do. And do you?

                    CUTTER
          I do. And do you agree to the
          standard rules and regulations.

Computer ANNOUNCES, "Nitrogen levels high."

                    KITTY
          So romantic. No way.

                    CUTTER
          I do. I pronounce you...I mean
          us, Husband and Wife.

Cutter takes his Naval Academy ring off and puts it on
Kitty's finger.

>   CUTTER (cont'd)
>   Listen to that.

>   KITTY
>   KISS ME.

Through the speaker we can hear the USS Leviathan making
MOANING NOISES like a moose.

>   CUTTER
>   What's that?

All of a sudden the controls in the pod move...

They start to ascend.

>   KITTY
>   We're moving up.

They both contemplate this moment.

>   CUTTER
>   Splendid.

>   KITTY
>   We're still...alive.

>   CUTTER
>   I knew it all the time.

Kitty looks at her finger with Cutter's ring, then they
slowly melt into each others' arms. A moment of tenderness
between them.

EXT. USS LEVIATHAN - DAY

The USS Leviathan is now entirely frozen and gets caught
right in the center of the bottom of the Whirlpool blocking
the flow, like a cork in a bottle.

EXT. BOTTOM OF THE OCEAN - DAY

We see the sub fracture and BURST into millions of frozen
pieces. It is a spectacular show of molten magma, liquid
nitrogen and cold oceanic waters meeting.

The Whirlpool slowly begins to close up.

INT. SINGLE PERSON ESCAPE POD - DAY

They slowly try to move around. The escape pod is caught in
the current and takes an irregular path.

>   CUTTER
>   It looks like the Maelstrom
>   Whirlpool is stopping.

The escape pod is moving upwards.

                    KITTY
          At least for now. But the planet
          is getting warmer, the sea levels
          are rising and we have a hundred
          times as many severe storms as we
          did fifty years ago.

Kitty is caressing him. The escape pod continues upward. We
see the depth indicator.

                    CUTTER
          Okay, but I suggest we continue
          with our plans from earlier now
          that the world is safe once
          again.

A long pause as the escape pod surfaces.

                    KITTY
          Is it? And about that engagement
          party. Did you ever pick up that
          suit?

                    CUTTER
          Yes.

                    KITTY
          Good. I accept your apology.

EXT. COAST GUARD VESSEL - DAY

We see the Coast Guard in the distance. The boat has all the
rescued people on it, saved from earlier.

INT. SINGLE PERSON ESCAPE POD - DAY

                    CUTTER
          We're at the surface at just the
          right time.

Cutter blows open the hatch. Kitty lets out Mister.

                    CUTTER (cont'd)
          The Coast Guard! And what a
          beautiful calm ocean.

                    KITTY
          At least for the moment.

                    CUTTER
          Kitty.

Mister barks. They LAUGH.

                                        FADE TO BLACK

THE END